The Essence of Life

By

Bee G. Gwynn, M.D.

Library of Congress Control Number: 2011918247
ISBN: Softcover 978-1-4653-7858-3
Hardcover 978-1-4653-7859-0

To order additional copies of this book, contact:
Xlibris Corporation
1-888-795-4274
www.Xlibris.com
Orders@Xlibris.com

Dedication

This book is dedicated to all the homeless and mistreated cats, dogs and horses.

All proceeds from the sale of this book will be contributed to the Humane Society of Davie County for the upkeep and care of all the animals in the adoption center. Any contributions to the Humane Society would be appreciated by the author and the hundreds of animals taken care of by the Humane Society of Davie County.

Humane Society of Davie County

291 Eaton Road

Mocksville, NC 27028

Table of Contents

Preface 5

Introduction and First Life 8

Ediacaran Period 12

Paleozoic Era

Ediacaran and Precambrian Periods 15

Cambrian Period 17

Ordovician Period 25

Silurian Period 29

Devonian Period 33

Carboniferous Period 37

Permian Period 39

Interval 42

Mesozoic Era

Triassic Period 48

Jurrasic Period 53

Cretaceous Period 58

Cenozonic Era

Tertiary Period, Eocene and Miocene Epochs 62

Quarternary Period 68

Human Evolution 70

The Dark and the Deep 76

Preface

This book is a synopsis of paleontology. It has not been written for scientists. Their contributions are herein utilized. It has been written primarily for the young, perhaps middle school to high school as well as for adults who are young at heart. Adults who have chosen other pathways for accruing a livelihood and who might just now begin to wonder how earth, life, all this, began and evolved. The work has been simplified in order to render it more likely to be read and understood by the non-scientist.

Chronological Scale

Epoch	Period		Era	Millions of Years Ago	Biological Development
Recent	Quaternary				
Pleistocene					First Humans
Pliocene					First Hominids
Miocene				26	First Apes
Oligocene	Tertiary		Cenozoic		First Monkeys
Eocene					First Whales
Paleocene				65	First Horses
	Cretaceous				Dinosaurs, Flowers
	Jurassic		Mesozoic	136	Dinosaurs, Reptiles, Birds
	Triassic			190	Flying Reptiles Oysters
	Permian			225	Reptiles Ginkgo
	Carboniferous	Pennsylvanian		280	Coal
		Mississippian		320	Insects
	Devonian			345	Fish
	Silurian			395	Dinoflagellates
	Ordovician		Paleozoic	430	Conodants
	Cambrian			500	Trilobite
	Precambrian				
	Ediacaran			570	2 Cell Layers

from Barnes, Invertebrate Biology, 6th Edition

O could I fly, I'd fly with thee!
We'd make with joyful wings,
Our annual visit o'er the globe,
Companions of the Spring.

"To the Cuckoo"

Michael Bruce

1740-1767

Introduction and First Life

In the beginning, there was a void followed by a tremendous unfathomable explosion, Stephen Hawking's theory of the initiation of the universe, The Big Bang, the beginning of time-space over 13 billion years ago. Rocks, dust, atoms, compounds flew apart expanding into space. Even though there is now organization of a sort, stars, planets, comets, galaxies, dust, black holes, universes, there is yet expansion. Planet Earth was formed 4.5 billion years ago by the sticking together of rocks as they collided and went through a series of volcanic eruptions, hits by comets, debris and other objects, its shape rounded, its hot core remained central and a mantle and crust were formed. Noxious gases were eventually expelled, hydrogen and oxygen combined in the atmosphere and water condensed and fell upon the Earth until it covered 71% of the Earth's surface. The Earth, with its water, cooled eventually over more millions of years. Initially, there were many landmasses. Earth was lucky-can one call it else but luck? It was located far enough away from our Sun, a ball of hydrogen-helium fusion, so that the climate favors life. Our sister planets (Mercury and Venus) are too hot and (Jupiter, Mars, Saturn, and Uranus and Neptune) too cold for life to begin or survive. In past years, it was believed that all life on earth was created by the Sun's action on green plants, a process called photosynthesis. Now, however, we are aware of life at the bottom of the ocean as well as in the dark, deep water column. Chemosynthesis as well as adaptation to darkness has resulted in life without the Sun's benefit. After the ozone layer formed in our atmosphere, Earth was protected from harmful radiation from the Sun, and life could begin here. The land very gradually broke up and over millions of years; it split apart moving into the many parts of the globe we now know. The water was always contiguous but separated enough so that oceans could be named. Thus, the Atlantic, Pacific, Indian, Southern and Arctic Oceans were named but remained continuous so that it was possible for life that formed to move from one ocean to another.

The origin of life, about 4000 million years ago, was preceded by a host of complex chemical reactions. The Earth's core consists of a liquid center of hot iron and nickel, the semi-liquid mantle with upwellings of volcanic lava forming the crust above which is composed of 71% sea water and 29% land. The atmosphere consisted of water vapor, nitrogen, carbon dioxide and small amounts of hydrogen, methane, ammonia, hydrogen sulfide and carbon monoxide. It was very hot and meteorites and comets hit the Earth perhaps bringing organic chemicals. It was these organic chemicals that made up the original broth that formed organic molecules of amino acids that bonded into protein. DNA and RNA were the first protocell. First life occurred when DNA and RNA began replication. First life was probably a virus which has no nucleus or cell membrane. A bacterial cell has a cell membrane. This form of life bounced around for millions of years. The first identified bacteria was cyanobacteria, capable of photosynthesis. From the hydrothermal vents at the earth's crusts we now know chemosynthesis takes place. This may have been first life i.e., bacteria that produce carbohydrates by chemosynthesis and establish communities of life forms (animals) on the earth's crust.

Most scientists believe bacteria and possibly, Archaea, a new domain recently found, were present as long as 4000 million years ago. This primitive protoplasm, Archaea, emits methane and can be found at Yellowstone National Park, as well as on the ocean floor. There were low levels of oxygen, so we know photosynthesis, a process in which water and CO2 with sunlight yield oxygen and carbohydrates (food) was taking place and now believe it came from cyanobacteria, which contain chlorophyll making them suitable for oxygen production. Oxygen percolated through the ocean and up into the atmosphere, but it took millions of years for oxygen to reach the levels we now enjoy. Cyanobacteria have no nucleus and are thought to live in mats (layer on layer) on the ocean floor and eventually on land as a crust. These mats are called stromatolites and can be seen in various parts of the world on shorelines, particularly in Australia. So what is it that makes bacteria a life form? In 1954, Watson and Crick found the helical structure of genes, DNA, or deoxyribonucleic acid and its similar form RNA, ribonucleic acid. It was responsible for the protein formation of all living plants, fungi and animals. Many scientists have tried to make a living cell by duplicating the process they think happened by making a brew of saline and other compounds and charged it with electricity, but alas, to no avail. The gene responsible for the protein required for life is self-perpetuating. The bases that connect the helical curl are adenine, cytocine, guanine and thymine. Usually, only the first letters are used thus, ACGT. Their formation, turned on or off, is of vital importance in creating life as we know it now and implies their presence which were life forms in discussing fossils. Darwin traveled all over the Earth collecting samples of animal and plant life and came to realize there had been an evolution of life by natural selection. It was not always successful in producing a strong species.

If a weaker one occurred it would die out and the natural selection that was successful was stronger and lived on. An example of this can be seen in the beaks of finches in the Galapagos, each beak suited better for the type of seed it fed on. Also, another of Darwin's interesting finds, that of the *Angraecum Sespidale*, an epiphytic orchid which has a long spur wherein lies the nectar. The pollinia are located at the beginning of the spur, so that as the predicted (by Darwin) animal that sought the nectar would someday be found with a proboscis long enough to reach the nectar and thus take the pollinia to another plant and fertilize it. Such an animal was found later on, a moth which has a proboscis 30 inches long, the spur of the plant 27 inches! This orchid is commonly called Darwin's Orchid.

The true condition of the sea when life began is unknown. There were probably low dissolved oxygen and a different combination of chemicals from those present now. There is a chronological scale of time as millions of years passed and what life existed at that time. As scientists learn more, these time zones are constantly changing. The oldest bacteria fossils were found and radiocarbon dated about 3000 million years ago. There were most likely many different bacteria that came and went in the ocean as the Earth was still being bombarded by meteorites. Organisms with hard tissues like shells and bones favored fossilization, but surprisingly many soft organisms left imprints on hard surfaces such as limestone, shale and sandstone. A cell must have organelles, nucleus and an outer membrane. This alone could be imprinted on a hard surface after its death; however, movement of layers of earth above due to earthquakes, glaciations or chemical changes could obliterate it.

Fossils are difficult to find and may even be in dangerous locations. Fortunately, the Earth spins on its axis and is therefore not subjected to heat of long duration. Seasons are reversed from north to south.

Recently life has been found on the ocean floor. There are black smokers like stacks of chimneys belching black sulfurous smoke, poking through the Earth's crust, a hellish sight, but feeding life all around, tubeworms, crabs, clams, sea stars and snails, a community of life now thought to have perhaps existed before life in sunlight. The energy of the smokers comes from volcanic action. All other energy comes from the sun. The smokers emit sulfurous gases. Living among the above listed animals are sulfur bacteria which convert hydrogen sulfide to carbohydrates and sulfur. This is chemosynthesis as opposed to photosynthesis which produces carbohydrates and oxygen but needs the stimulation of sunlight. The bacteria produce food for these animals deep in total darkness of the seas and under huge pressure. But there they are, a community of life of a strange essence. Another instance of strangeness is the recently found cold seeps on the ocean floor. There are many types of feeding going on here: chemosynthesis, methanogenesis, and strangely enough a bivalve attached to the egg emitting portion of a female tubeworm. Who said life is not sneaky as well as diverse? There are also methane oxidizing bacteria. White crabs and mussels live among them. There are also other chemosynthetic communities on the ocean floor called cold seeps. These were discovered in the last 30 years. They seep hydrocarbons and methogens at high or low temperatures. Usually they support only microbial, i.e., cyanobacteria mats but many other life forms are present such as mussels, gorgonians and corals. Vent sites on the ocean floor teem with a wide array of life such as tubeworms and huge clams in the Pacific Ocean and eyeless shrimp in the Atlantic Ocean floor. Hot, rich mineral water from the earth's crust rises upward and teems with a wide array of life. "Old Faithful" at Yellowstone National Park is such a vent; a connection from the Earth's crust to the Earth's surface. Archaeans, single cell prokaryotes, have been found here since the 1950's. These creatures look a lot like bacteria, but chemically and genetically they are different and have been categorized into their own Domain. Recent research has found Archaea in the ocean plankton. There are hot vents and springs off the coast of Ecuador. The temperature is 1000 degrees centigrade and here growing placidly are great tubeworms and huge clams and mussels. Chemosynthesis supplies the food. Scientists are working now in order to ascertain the origin of life on earth----ocean floor or earth's surface?

The classification of all living things is as follows:

Archaea Bacteria	No nucleus

Plants Fungi Animals	Eucharyotes (have nuclei and higher development)

Fungi are known to occur in the ocean but they have not been well studied.

Ediacaran Period

The Ediacaran Period (900-800 Million years ago) has fragile fossils mostly found in sandstone. They were multi cellular and soft bodied and seem to have no middle layer. Their shapes are square, quilted and some nearly blobs. None lend themselves to evolution and have been called a "false start".

Ediacaran Imprint

They have been found all over the globe, but more so on the Ediacaran Hills of South Australia. It was the creation of the cell with a nucleus, organelles (inclusions such as chloroplasts and mitochondria, essential for metabolism) that separated simple life forms from life forms, capable of evolution. Many one-cell life forms, protists, remained as one cell but developed different shapes and capabilities. Some protists became part of the plankton which literally floats on the ocean's surface. These are phytoplankton or one-celled algae such as diatoms, dinoflagellates, foraminifera, radiolarians and coccoliths. Phytoplankton fossils have been found through almost all periods changing and increasing in diversity.

Diatoms

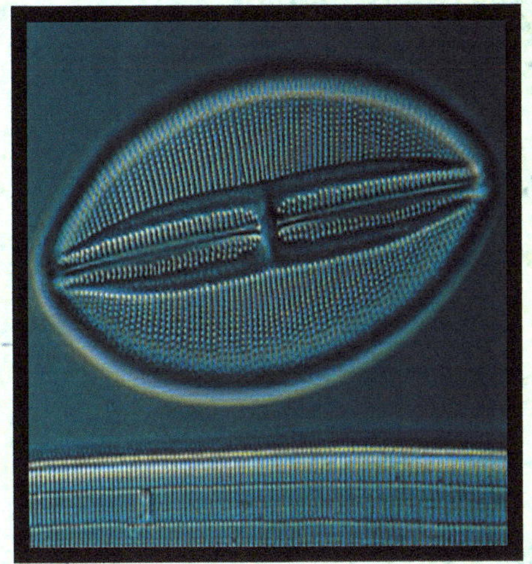

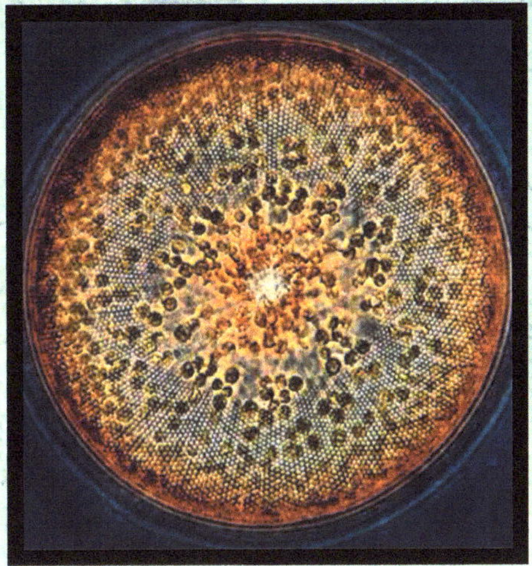

Coccolith

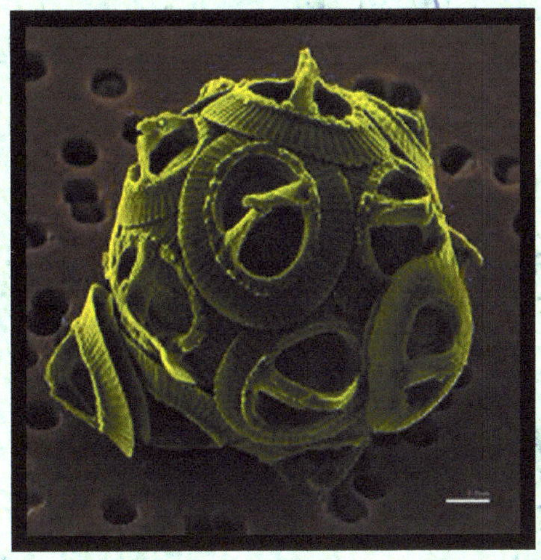

The same may be true for the appearance of zooplankton which is composed of larval forms of sea animals such as crabs, mollusks and echinoderms (sea stars, brittle stars, urchins and sand dollars). Both types of plankton float intermixed on the top lighted layer of the oceans. This is the beginning of the food web in the oceans. Where there is up welling of nutrients (nitrogen, phosphorus, silica and more from the ocean floor), the plankton thrives and fisheries thrive. Another important member of plankton is the copepod, a zooplankton that is multicellular and classified as a crustacean.

Copepod

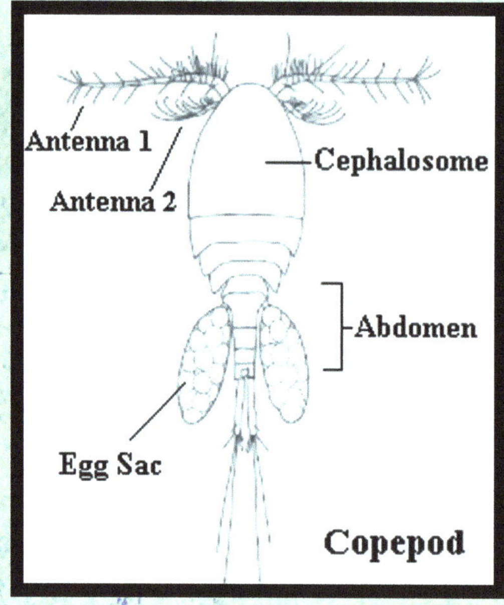

There are many sizes and species of copepods, each is said to be the perfect size for the mouth of a recently hatched fish. Another member of the plankton most important for all life is the diatom of which there are also many species. Studied microscopically diatoms are the most beautiful and essential members of the plankton family. A phytoplankton, a diatom has a double silica shell which fits together as a capsule. Inside, diatoms are chocked full of chloroplasts containing the important chlorophyll with barely room for its nucleus. Phytoplankton are high in the ocean in order to catch every beam of sunlight possible. The chlorophyll inside the chloroplasts, an organelle, undergoes photosynthesis, the product of which is carbohydrates (sugars) and oxygen. There are two chemical reactions that take place, 6 molecules of carbon dioxide plus 6 molecules of water yields sugar and oxygen, the light reaction (where sunlight must be present in order to activate an electron from hydrogen which is then transferred to the carbon). In dark reaction, or Calvin Cycle meaning light, although present, is not necessary, the carbon undergoes what is commonly called "the sugar shuffle", creating carbohydrates, food. The by-product of the reaction is oxygen. The volume of oxygen produced by phytoplankton is thought to be greater than 50% of all the oxygen in our atmosphere. They become more important as we continue to decimate trees and other foliage plants on earth in the name of "development". Is it development or destruction? We must protect the oceans from pollution if we are to rely on phytoplankton for the oxygen all animals need to survive.

THE EDIACARAN AND PRECAMBRIAN PERIODS

These years meld together. There are hints of future things to come such as the sudden appearance of the spicules. Spicules are tiny objects with three or more points that hold a sponge together. The sponges we know now may have any of three types of spicules-glass, spongin or calcium. These were left behind as sponges deteriorated, their fossils were found, and carbon dated. The choanoflagette, a one-cell animal, has many flagella that appear as a kind of tail.

Choanoflagellates

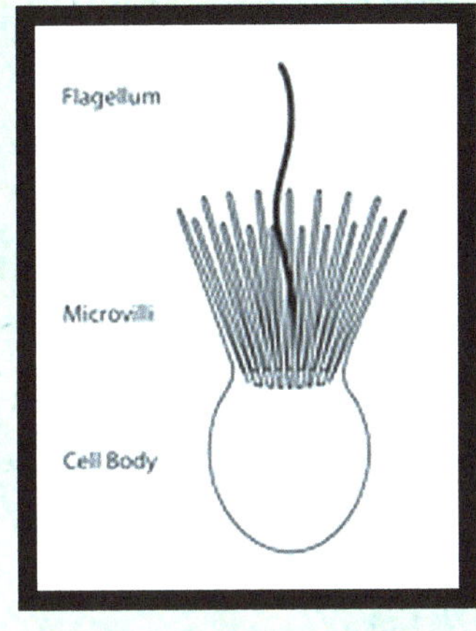

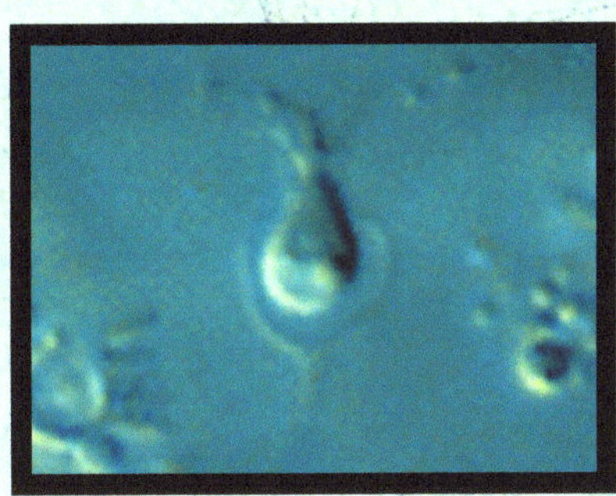

The sponge is a collection or conglomeration of choanoflagellates studded with pores that let seawater with its nourishment travel through and waste out of the sponges. Fossils of small sponges have been found in the Ediacaran and Precambrian eras, 800-500 million years ago.

Fossils of jellyfish have been found scattered around the globe during the Ediacaran and Precambrian years. Their radial symmetry, two layers of cells and lack of organization point to their primal arise. They are all gonads, stomach and tentacles with the one new possession, the nematocysts which contain stinging cells. Relatives of these early jellyfish remain with us today and are to be avoided. If you see a lovely purple balloon on the beach, don't touch. It's probably a Portuguese-man-o-war and stings after death. Find a stick and bury it on a dune to protect some unknowing child. At least jellyfish we can come to terms with because they survived and evolved. The other Ediacaran-Precambrian fauna remain curious. Bacterial mats of cyanobacteria were still intact, not preyed upon, yet. The whole of the Precambrian, i.e., from Earth's formation to the beginning of the Cambrian period is about 85% of geologic time.

The proof of the fact of evolution lies in the tool of Carbon dating. Carbon is a well known and plentiful atom. It has 3 isotopes, meaning the number of neutrons and protons are changeable. As it decays C14 becomes C12 and the ratio of the two can be used to calculate the amount of time that has passed since that creature died. Now this process has been revolutionized by a technique called mass spectrometry.

THE CAMBRIAN PERIOD

About 350 million years ago and over the next 70-80 million years, there occurred a rapid and curious increase in the amount and diversity of life on the planet Earth. Many of the fossils of these animals are the same or similar to life we now know in the ocean, however many others have gone extinct and their fossils are very curious, maybe even strange. The increase in the amount of life found in this period has been attributed to a rise in the amount of dissolved oxygen available, an increase in the amount of food to prey on, and water salinity more propitious to life as well as a decrease in noxious elements in the water and many other factors. We do know that multi-cellular life was on the rise and many animals had shells. The phyla, as we know them today, were mostly formed in the Cambrian Period. Triploblastic life (meaning there were three layers of cells in the development) began in most animals, except sponges and jellyfish. Bilateralism had its beginning in the Cambrian, meaning there was a beginning of symmetry, or mirror image of each side of the body, such as our own. Fossils known as "small shelly fauna" have been found all over the world that date to the Cambrian, some are only 1-2mm long, but a beginning was made. The earliest trilobite dated 530 million years ago and many species were found concomitantly suggesting they had been around for some time. The trilobite is the "star" of the Cambrian Period. Books have been written about them alone.

The Cambrian explosion continued for 10 million years, a small amount of time when one considers the billions of years that had passed from the formation of the Earth and longer still from Hawking's Big Bang. One scientist in particular, Darwin, worried that there seemed to be no predecessors for the Cambrian animals. Having staked his reputation on the theory of Evolution, he had a difficult time coming to terms with the Cambrian. He was ill off and on during his life after the Beagle Voyage. No diagnosis was ever made which makes one wonder if he had doubts, not only because of religion, but also because of the Cambrian question. No predecessors of the Cambrian fauna have been found to date. Mental worries have their definite effects on the body itself. Most of Darwin's problems were centered on headaches and gastrointestinal problems both of which can be psychosomatic. His wife was very religious and they never spoke together of his book. Pride, however, ruled the day as it should have. When Darwin learned of Wallace's same theory of Evolution, he was quick to publish and his aches and pains decreased.

During the Cambrian Period, life had begun on land as if in preparation for land animals. Green algae, cyanobacteria crust began to develop waterproof coats to prevent desiccation. Mosses began to grow along with Sporangia (a structure containing spores) liverworts (a plant resembling moss), and bryophytes or non-flowering plants and have been dated to the mid to late Cambrian. By mid-Cambrian, there were filamentous algae, but most of the land was barren, with warm seas and polar ice absent. Back to the Cambrian Sea, after sponges and jellyfish had gone their merry way to evolve many fold, triploblasts (3 cell layers) began to appear in great numbers and with much variety. There were the "hard shellies" early on and later animals of great abundance. In addition to the trilobite that has already been mentioned, there appeared arthropods (crabs) and a host of other sea animals that have gone extinct in addition to the phyla (groups) that we now know in the oceans. The Burgess Shale is located in western Canada. Burgess was the name of a previous governor of Canada for whom the pass to the shale was named. Charles Walcott discovered the fossils of the Burgess Shale high in the Canadian Rockies of British Columbia, in 1909. These animals appeared all to have died at one time due to a geological shift of some kind. They had been buried so rapidly that even their soft parts did not decay. Stephen J. Gould, a well known paleontologist, wrote a book describing the fossils of the shale called Wonderful Life, highly recommended for the pictorials and descriptions of these strange creatures which are now protected at the Smithsonian Institution. Walcott died in 1910, but he had collected at least 65,000 specimens. He had hardly scratched the surface. The quarry was then explored by the Geological Survey of Canada. Research students working on their PhD's were signed up to dissect and draw the individual fossils. They were Harry Whittington who supervised the work, Derek Briggs and Simon Morris of the University of Cambridge. The fauna were more diverse and unusual than Walcott had realized. Many had bizarre anatomical features, unlike any creature that had been seen before-a weird essence of life. The names given these creatures were equally weird. Walcott was responsible for some of the names such as "Marrella" citing the fact that it looked rather like a lace-crab with rows of gills. Other names are thought to have been derived from nearby Indian tribal names. The first published paper by Harry Whittington was on Marrella in 1971.

Marrella

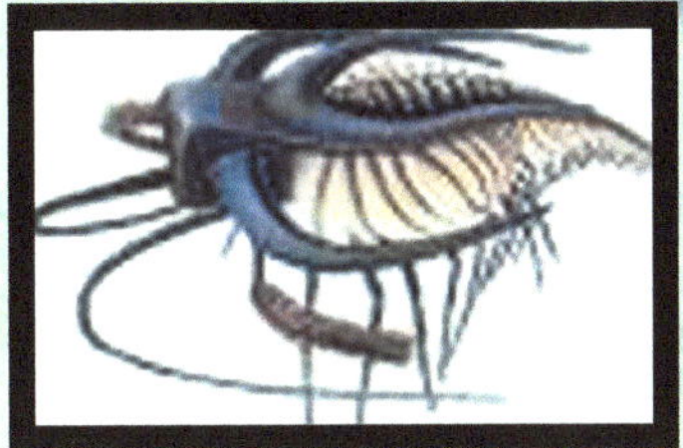

Later Whittington papers include "Anomalocaris" in 1985, "Wiwaxia" also in 1985 and "Sanctacaria" in 1988.

Anomalocaris

Wiwaxia

Sanctacaris

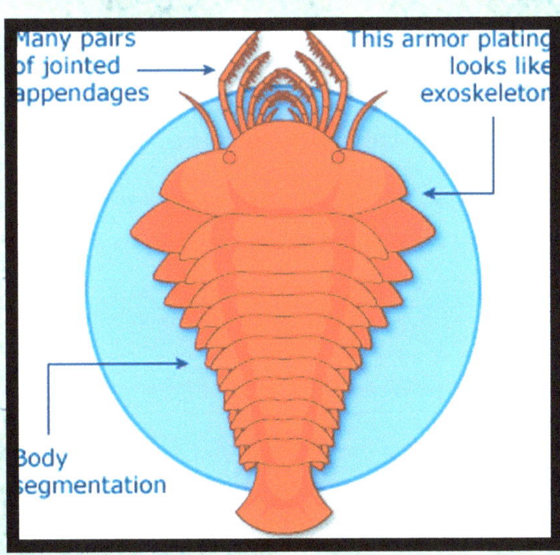

Sanctacaris is thought to have been a chelicerate that falls in the group of creatures including mites, spiders and Horseshoe Crabs. Horseshoe Crabs have long been called "living fossils", but perhaps this is giving them too much credit. They can be traced back only 20 million years, not the previously thought 200 million years. Be that as it may, they are common in the Atlantic Ocean crawling on the ocean bottom and creeping to shore to lay and fertilize their eggs on a high spring tide every year. While on the coast many of these crabs are taken by technicians who know how to draw their blood-which is blue due to copper (instead of iron like us) as the center atom of the hemoglobin. The Horseshoe Crabs are returned to the ocean after bloodletting unharmed. The blood is then spun in a centrifuge and the white cells at the bottom of the tube are freeze-dried and sent to laboratories all over the world. The white cells are then reconstituted and used to test all objects that are sold as "sterile", such as surgical tools. If the tested object is contaminated with life threatening gram-negative bacteria, it is not sold.

So who would have thought that the precursor to a not common but well-known living fossil could be so important to us?

Horseshoe Crab

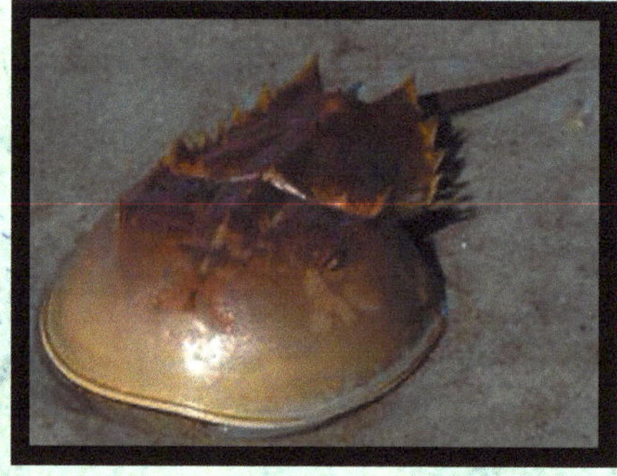

The trilobite is the best known, largest in quantity, and longest lasting of all the Burgess fauna and perhaps a precursor of the Horseshoe Crab. Trilobites lived until the large extinction in the Permian Period, about 248 million years ago, a span of about 250 million years.

Trilobites

Fossil **Drawing**

The trilobite was thought to be an arthropod (meaning jointed legs) but their legs were hard to find, having broken away from the body of the fossil in most cases. Its body was segmented. All pairs of legs were thought to be biramous although the gill branches were closely associated. The body consisted of a series of repeated segments, each segment usually having a pair of biramous (2 prong) limbs. Because trilobites lived so long they had plenty of time to undergo minor changes and differences, i.e. evolve. The first image forming lens to evolve probably belonged to a Trilobite. The lens was made of crystal calcite, as is the lens of the eye of the Brittle Star, an echinoderm. There are many more fascinating creatures from the Burgess Shale. Pictures, animation and groupings can easily be found on the Internet. You are urged to look further into this wonderful group of creatures that may have evolved into arthropods, even Chordates that we now know. You will be entertained as well as educated, which brings us to a description of other animals that arose in the Cambrian era. Jellyfish and sponges have been noted to have begun in the Precambrian Era. They are with us now in many forms and colors, especially the sponges. One of the jellies, the ctenophore or comb jelly does not have a stinger. It is a beautiful egg shaped oval usually pelagic (lives out to sea) creature with eight rows of cilia that sparkle with colors in sunlight. Ctenophores also have bioluminescence. In the dark of night, if touched, or jostled about by waves they will flash a greenish light. Many marine animals have this capacity, especially in the deep sea where there is no light. Luciferan, an organic chemical, as well as luciferase, the enzyme, must be present to create the photon of a heatless light. A stream of photons gives us light. Flatworms, round worms and tubeworms probably arose in the Cambrian and were prey for the Burgess Shale and other animals evolving. Reproduction had gradually moved from single cell mitosis (dividing) to sexual. Therefore, there was an exchange of genes by the time of the Cambrian. Bilateralism became the norm, as it is with us humans. The formation of coral reefs began in the late Cambrian. Corals are polyps with tentacles that can reach out for food particles. This activity however, primarily takes place at night. Coral polyps have dinoflagellates minus the flagellae, inside their cells, as do sea anemones. These are one-celled zooxanthellae that are symbionts (live together each helping the other). The polyp gives the zooxanthellae a home, and in exchange the zooxanthellae manufacture sugars for food via the process of photosynthesis because they contain chlorophyll. Stony coral polyps secrete calcium carbonate forming the beautiful coral reefs in warm water, therefore are near the equator and high in the seas in order to capture sun beams for the zooxanthellae. There are many species of stony coral polyps-brain corals, staghorn corals, elkhorn corals, plate corals and many more. Where there are coral reefs, a whole ecologic community exists. Many marine flora and fauna live in crevices or circulate around the reef. Scuba divers are drawn to these reefs for their beauty as well as their diversity of life.

Brachiopods appeared in the Cambrian as did crinoids which are stalked echinoderms, looking a bit like flowers with stems in the sea floor. Feather Stars were present in the Paleozoic also. The Paleozoic is an era that includes the Cambrian through the Permian periods. See chart on page 6.

Echinoderms were beginning to evolve i.e., sea stars, urchins, sea cucumbers, brittle stars (already mentioned for their eyes) and sand dollars. Sea stars also have a tiny eye at the end of each arm. They have great powers of regeneration. If they are harmed and have but the central disc and one arm intact, they will regenerate into a full five arm star. This may also be true for the many-armed stars found in the Pacific. The sea cucumber is interesting in that it has mutable connective tissue. It can be pulled apart so that only a string of tissue separates the two ends and over time, it will reconnect itself into a whole, unharmed sea cucumber again. The sea urchin has little pointed teeth on its under or oral side. It chews food as it crawls along the sea floor using its podia (feet) and above it is a digestive chamber called Aristotle's Lantern, named for the man who first described it. The sand dollar, which is green when alive, has a petal design on its aboral side. These are gill areas. On the oral side are food grooves and the teeth are inside the center opening. If it is found on the beach, it is white and dead and if you shake it you will hear the teeth rattle.

Other organisms that originated in the Cambrian seas include small mollusks that were the original "shelly fauna" and are now our clams, oysters and scallops or bivalves. Scallops have eyes and they are a beautiful blue. If you see a live scallop (while you are walking in a bay perhaps) pick it up and hold it perfectly still for a few moments. You will see its valves part slightly and there are 24 pairs of blue eyes looking at you.

Ammonites, which fall in the phylum Mollusk, originated in the Cambrian and were found also in the Silurian period. Ammonites went extinct in the asteroid collision with Earth that killed all the dinosaurs. Nautaloids survived or evolved later on. The Nautilus is now found in the Indo-western Pacific Ocean. They have a coiled shell. The inner part is separated into compartments by septa. These are empty gas filled chambers; the animal lives in the outermost chamber, open to the sea from which it grabs food with its tentacles.

Other mollusks made a beginning in the Cambrian, but it took several more million years to evolve into the gastropods we know today such as spiral forms, whelks, conchs, cones and many others.

Brachiopods are thought to have arisen in the Cambrian as well as a possible chordate, meaning it has a notochord, the precursor of the vertebral column. Brachiopods look a bit like bivalves but are very different. Brachiopods, commonly called lamp shells, have 2 valves but they are unequal leaving an opening for sea water to flow through. Their inner structure is adapted for filter feeding.

One of the Burgess Shale fauna is thought to perhaps have had a notochord (precursor to vertebrates). Its name is Pikaia and it was swimmer.

Pikaia

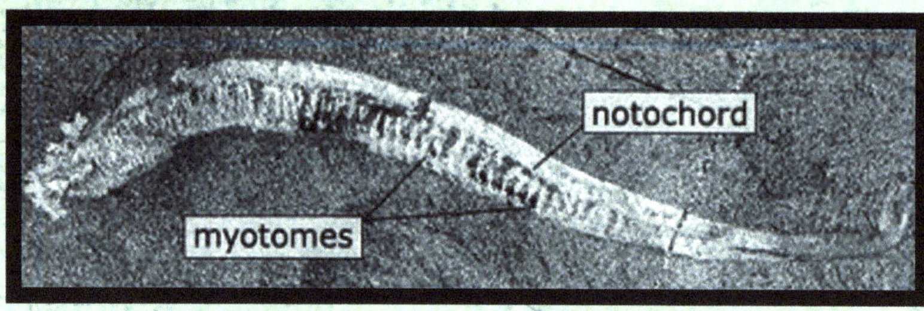

Its fossils were recently discovered and believed to be a chordate-as are fish, dinosaurs and humans!

The Ordovician Period

Before going forward, a few items need explaining. The common names of animals and plants are being used here in order to simplify understanding. In a book written by a scientist, the Linnaeus system of binomial nomenclature would be useful perhaps in addition to the common names. You would know the genus and species immediately, therefore if you were speaking with a scientist from China and he or she knew no English, he would immediately know what creatures you were speaking of. In addition, if we were true scientists, we would immediately know which era and period of time that plant or animal existed or was first discovered because we had memorized the Chronological Chart. The late Stephen Gould advised his students to use a mnemonic and make it naughty so one remembered!

The Ordovician Period is described as existing from 488-443 million years ago. It follows the Cambrian and precedes the Silurian. It started with an extinction event that lasted about 44 million years and exterminated about 60% of marine life. Some of our old friends that delighted us so in the Cambrian will not be seen again, having gone extinct. Remember that the Earth was still in an unstable state. Volcanoes, asteroid hits and tsunamis were common occurrences. The atmosphere was hot, sea levels were high and massive amounts of carbon dioxide turned the planet into a hot house. Calcium carbonate production increased so that new invertebrates could have a rigid exterior, good for protection. Sea levels rose and then fell. Trilobites continued to increase in numbers and species and because of high sea levels, fossils have been found on Mount Everest as well as the Rocky Mountains. New creatures include graptolites, conodants, brachiopods and the first vertebrates, which were jawless fish, the hagfish, which was probably a filter feeder at this stage. Later on hagfish are known to invade dead bodies of whales and large fish and literally suck the dead flesh into their jawless but greedy mouths. The graptolites were a ragtag assemblage of the plankton, now extinct. They were colonial animals with strange shapes. Their fossils are varied in shape and some have been found to have tiny tubes inside the structure. Some are called hacksaw fossils.

Graptolites

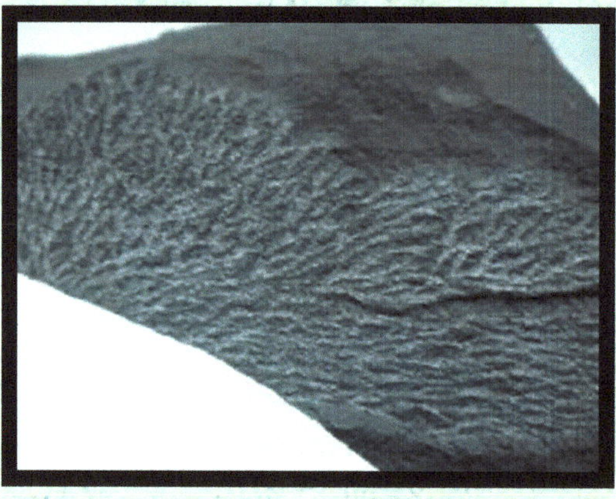

They are index fossils that allow carbon dating for the Paleozoic rocks.

Conodonts have an interesting story. The fossils were hard and made of calcium phosphate. They were long thought to be the teeth of some creature, but in the 1980's, it was determined to be a vertebrate perhaps related to Pikaia of the Burgess Shale. It was then found that the conodont had a pair of bulbous eyes and was at last identified as a fry (a fish stage of embryological development) of a Salmon! "But, you say, Salmon had not evolved in the Ordovician". And you would be right!

Conodont

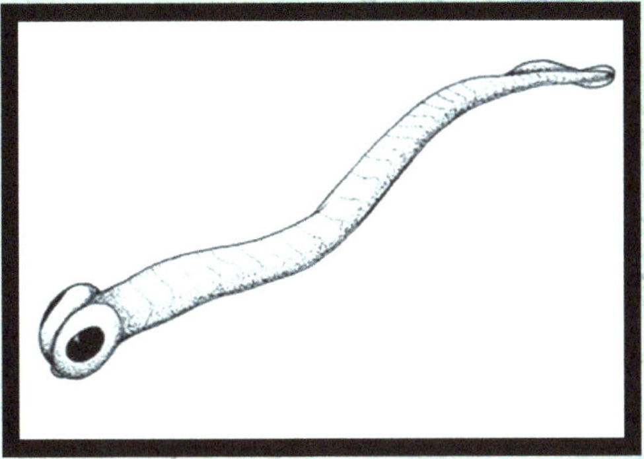

So it retained its name and is thought now, having been studied under the microscope and the muscle bars and fins clearly seen, that it was an independent animal in its own right. The conodont puzzle was finally solved. They were a type of early fish with vertebrae, thus distantly related to humans. However, another group of paleontologists does not believe it is a vertebrate and classifies it as a type of worm. You pay your money and you take your choice-old Chinese proverb.

There are niggling differences of opinion throughout the study of Paleontology; someday the truth will unravel. Bryozoans made their appearance in the Ordovician. Both bryozoans and brachiopods are filter feeders, however bryozoans have soft bodies while brachiopods (both extant) have two unequal shell halves. Their two valves were anchored to the ocean floor and seawater flowed through the two valves, protective of the filtering mechanism. Corals, both stony and soft were present. Reproduction of stony corals is much like that of sea urchins, except that the corals do no traveling. They are fixed in their reef. On a full moon, and as if orchestrated, they release eggs and sperm that combine in the sea water and float off, hopefully to start a new reef. Sea urchins have to travel on the ocean floor to come together, then release eggs and sperm.

There was an increase in the numbers and species of filter feeders during this period. Brachiopods and stony corals were more numerous than trilobites caused by the increase in available calcium carbonate.

Bivalves, gastropods such as conchs and cones and some early ammonites appeared. Ammonites and Nautaloids arose similarly, but Ammonites are extinct and Nautaloids evolved into the extant Nautilus.

Reef corals have been present since the Cambrian. They have been torn apart by typhoons and hurricanes, but regenerate themselves afterward. The ecosystem that surrounds them also returns as if drawn by magnetism. Before 230 million years ago, coral reefs were different from those we have today. They were inhabited by sponges and bryozoans, small soft bodied filter feeders, because there was an abundance of these. As millions of years passed, droves of new species began to inhabit the coral reefs and they grew to great sizes. Think of the Great Barrier Reef of Australia seen from space, to the small Pacific atolls no larger than a small town. Worms tend to be overlooked in science, largely because they burrow underground or under the sea floor. Velvet worms which survive today under rotting logs, were acutally evolved from a stumpy legged animal of the Cambrian. On land there were green algae, mosses and liverworts and a few fossilized hyphae of fungi.

Liverworts

Liverworts are non vascular plants with deep lobed leaves found in humid locations to this day. Imagine a stark scene of rocks of various sizes and small patches of green scattered around sites of moisture. It was not a pretty sight. It was desolate and silent. Even though ozone had been formed in the atmosphere protecting Earth from the sun's harmful rays, the temperatures were erratic and at the end of the Ordovician, there occurred a glacial event followed by extinctions.

The Silurian Period

The Silurian Period extended from the end of the Ordovician to the beginning of the Devonian, that is 443 million years ago to about 416 million years ago. It endured for 44 million years and began after the glacial extinction period at the end of the Ordovician. Do be aware that these dates are not exact and are constantly changing as paleontologists acquire more knowledge. The Silurian period enjoyed warm temperatures and no glacial events, glaciers left from the Ordovician rapidly receded. This period enjoyed a relative stabilization of climate. The first bony fish evolved covered with scales. It had gills and movable jaws. Huge sea scorpions appeared, trilobites continued to diversify and leeches made their appearance. Parasites have always been with us. As they develop they loose the part of their anatomy they no londer need. All parasites really need is an attachment, a gut, and a reproductive system. They appeared in the Ordovician and are with us still, in the ocean, on land, on plants and on all animals. Brachiopods, mollusks, trilobites and arthropods (crabs) were plentiful and diverse. The coral reefs became stronger and larger and the new fish flocked to them. The plankton were increased in numbers and variety. Phytoplankton became more anatomically specious (varied) and zooplankton (animal) became more diverse as bottom dwellers evolved in variety and species. Diatoms, one celled algae, covered with a capsule of silica, were more numerous. Diatoms along with other top dwelling algae continued photosynthesis adding more and more oxygen to the atmosphere as well as to the ocean water. One particular one-cell algae, the dinoflagellate (it has two flagella) made its appearance in Silurian. This particular algae is extremely interesting to study. Alive under the microscope it moves in a jerky manner due to the placement of its 2 flagellae, one around the girdle and one up and down the sulcus or groove.

Dinoflagellate

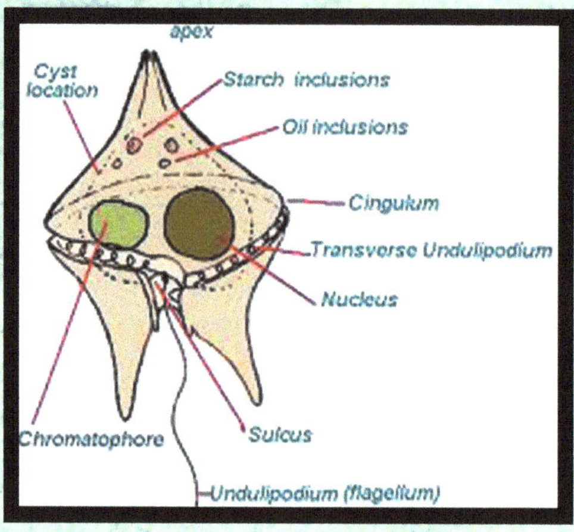

They were first described from core samples later discovering the living forms. The zooxanthellae, symbiots of stony corals, were not found in the fossil record. The chloroplasts (which contain chlorophyll) in most photosynthetic dinoflagellates are thought to have been ingested algae, however some have chloroplasts with different pigmentation and structure suggesting their chloroplasts were incorporated by other methods. The dinoflagellate is a cell of extreme complexity and capacity. It contains organelles, as well as a light sensitive eye or stigma. Some species contain toxins that cause fish kills (red tide) and ciguatera food poisining.

Coralites arose called tabulate corals. As the name implies, it was shaped like a table top and lived in warm, shallow seas. A cross section looked like a honey-comb. The brachiopods developed concentric and radial ribs on their shells similar to the markings on contemporary mollusks. Small bivalves clung together or fossils were found en masse. Some had byssal threads presumably holding them on the ocean floor. Crinoids, sea lilies, (echinoderms like sea stars and urchins) were shaped like a vase with a stem attached to the ocean floor and the vase upward.

Crinoid

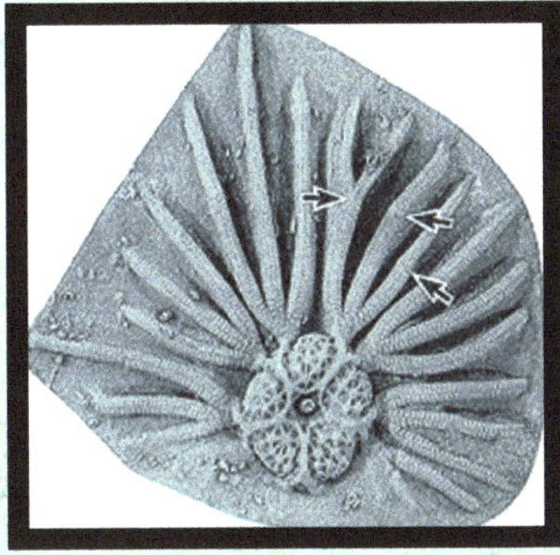

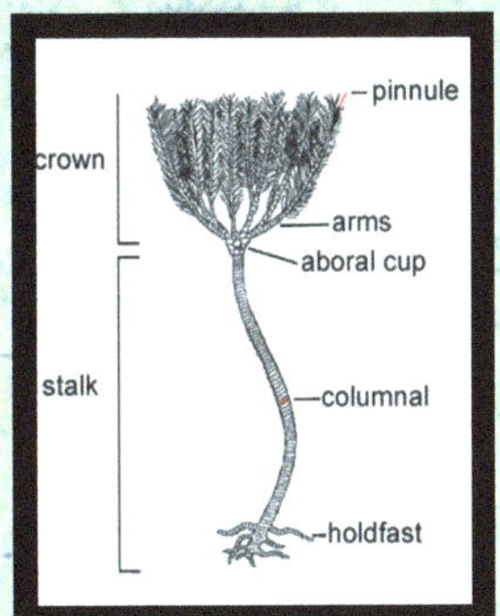

There were arms like tentacles used to grab food particles from the sea water. Crinoids are thought to be the oldest form of echinoderm with the oral aspect facing upward. The free moving echinoderms which followed all have the mouth facing down so they feed as they glide across the sea floor.

The Silurian sea was home to the first bony fishes and the development of the all important jaw and gills, the importance of the jaw, especially with teeth, is that the fish could now grasp and chew instead of just sucking as with the earlier hagfish, although hagfish are still with us today. Gills were adapted to filter oxygen out of sea water as it flowed over them. The thelodont was an early flat fish that was a bottom feeder, sifting through mud. It had a forked tail.

Thelodont

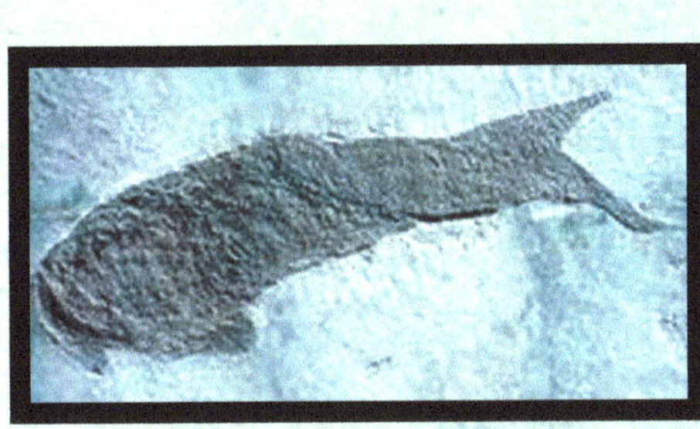

We have seen a few mosses on land during the Ordovician, but the Silurian saw the first vascular plant, Cooksonia.

Cooksonia

It was a primitive plant with xylem and phloem but no differentiation in root, stem or leaf. It did possess stomata which are pore openings for gas transfer. Inflowing carbon dioxide and outflowing oxygen resulted in the process of photosynthesis.

There is some evidence of the presence of spiders in the late Silurian. Even a sea scorpion ten feet long was found among Silurian fossils. Mosses continued to grow on land along with the *Cooksonia,* so as one looks out across the Silurian plain, there is more green and even a little suggestion of earth here and there. This new soil on earth would be due to the rotting of old plants. There was still silence and little to see on land.

The Devonian Period

This period has often been called the "age of the fish". The Devonian followed the Silurian and its timing is from 420 million years ago to 350 million years ago. Many species of fish (with scales and teeth) evolved and plant diversification was equally significant. It was in the late Devonian that fish began to come on land, fish that had bones in their fins and were marked to become tetrapods (four legs). In addition, the continents were on the move, plate tectonics were beginning and the climate became favorable for evolution and growth.

The coelacanth or lobe finned fish lived deep in the ocean. They were thought to be extinct until 1953 when one was caught in the Indian Ocean and taken to scientists for examination. Its four lobed fins were found to contain bones, not articulated, but in the general arrangement of human bones of the extremities. Recently, in 2011, divers photographed these Coelacanths at great depths and noted that their lobed fins worked in alternation, such as a tetrapod walking. Somehow tetrapods were "thought of" about 300 million years ago when their fossils were identified. Many other species of fish evolved during this period including the *Tiktaalik.* It showed even more tetrapod features including a series of bones that joined the skull to the shoulder girdle.

Before we make landfall and delve deeper into the Devonian, one should have a clear understanding of the evolution of the vertebrate column itself. In the study of Invertebrate Biology, i.e. from one cell animals to multicellular animals, there is an increasing diversity of cellular construction. Think of the sponges early on and arthropods later, and there in the Devonian we have the appearance of fishes with vertebrae. "From where did the vertebrae come?", you ask. It was certainly represented in the Cambrian by the Pikaia with its distinct notocord (nerve cord) running down the center of its body. The amphioxus also has a true notocord and because it is extant (living), it can be studied. The amphioxus is 5 mm long and lives in the bottom of the ocean with its filter feeding anterior and gill arches in the sea water and the rest of the body in the sand. It is the rest of the body that contains the notocord. It is a hollowed out nerve cord which contains a jelly-like substance. In the vertebrae of those that possess one, such as humans, the jelly-like substance of the notochord has thickened and becomes the inter-vertebral disks.

During the Devonian, sharks and their relatives, rays and skates, began to evolve; sharks quickly rose to the top of the food chain, but interesting enough they (nor the rays and skates) are not vertebrates. These animals have cartilagenous back "bones". It allows them to make quick turns in order to catch prey. Their skin is covered with hard denticles that afford them protection from an onslaught.

Which fish was first to make landfall? There are several nominees. The coelocanth can pretty well be ruled out as their fossils were never seen near shore and they live now in deep water. Not only did the lungfish have some lung, it also had a bone that attached to the shoulder, a bone similar to the human humerus. With its lungs, it could certainly go back and forth between land and water as the amphibians do. Frogs live on land but their tadpoles need the water. The missing link could well have been either the lungfish or the *Tiktaalik.* About this time (when theses fossils were being studied) cladistics reared its ugly head and scientists went berserk. Cladistics is a system of taxonomy (classification) in which the animals are all related to a common ancestor. The Cladists talked to each other and the scientists to each other and never the twain should meet. There was an uproar. As it turned out the Cladists were probably right. Cladistics is mainly founded on the anatomy of the animal in question and from its similarity to its common ancestor. Another way of analyzing animal relationships is by the study of its molecular structure. Molecular biologists study the sequences of proteins and project a common ancestor. All acknowledge that the first tetrapod on land was the Ichthyostega.

Ichthyostega

It was about 5 feet long and its fossils were first found in Greenland. It has ribs and paddle like hind feet, so that it could hunt in water or on land. Bryozoa, brachiopods, trilobites, crinoids, corals and other invertebrates continued to thrive and evolve. Early squids made their appearance along with arthropods.

Devonian plants were increasing vastly. The Earth was greening. There was more green algae, mosses, early forms including Horsetails.

Horsetails

Vascular plants evolved and trees began to grow because now nutrients and water could be carried from the roots to the leaves at the top. There was no branching. Because the roots were shallow (not much soil had built up yet) and the trees were top heavy, many would collapse if the leaves increased too much in size. There was a lot of photosynthesis going on. In the middle Devonian, branching began and primitive seed plants began. Spores have no outer protection. Seeds were well protected and an important step toward reproduction, less dependent on water. The Devonian landscape was dominated by Prototaxites Logani that grew 26 feet high.

Prototaxites Logani

It is one of the most enigmatic plants known. They had a bark like trunk and the internal structure had concentric growth rings. Recently it was finally found to be the fruiting body of a fungus. There were some plants with creeping rhizomes, low branching plants. The *Archaeopteria* tree in the late Devonian had frond-like leaves and grew to 26 feet with a 5 foot diameter. It produced spores (like ferns). The 3 foot *Elkinsia* is the earliest plant known to produce seeds.

Elkinsia

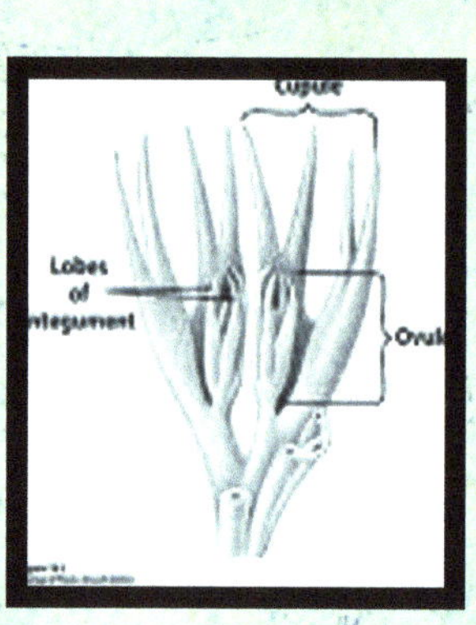

Seeds have a capsule that protects them and renders them more long lasting. It is a very important turning point in plant evolution. Now in your mind try to picture the Devonian landscapes with its spore bearing trees, low plants and fungi cones-a very different picture from the Silurian, **but still there is silence**.

The Carboniferous Period

This period took place from 359 million years ago to 299 million years ago. The name comes from the Latin for coal, *carbo.* Many coal beds were produced during this Period around the globe. This Period is usually broken into two parts, the Mississippian (earlier) and the Pennsylvanian (later). The land mass, Gondwandaland, is coming apart as plate tectonics dictate. Active mountain building took place and a drop in sea level caused a marine extinction, especially hurting crinoids and ammonites. This period saw the first insects go airborne. A dragonfly could be as big as a bird. Winged insects became diverse, finding that food was easier to see and sex was possible while airborne. The lycopods made a surge of growth, many growing into large trunks with heights to 80 or more feet. These trees were shallow rooted, had fern-like fronds for leaves, typical of lycopods, and vascular tubes in the trunk which carries nutrients and water to the leaves, even against gravity. As the leaves became dry, capillary action drew the fluids upward. The leaves, though fern-like had stomata (pores) and with their chlorophyll performed photosynthesis. When the trees collapsed, be it from age or top heaviness, they turned into coal if no oxygen got to them. The trees grew in dank, humid swamps, surely an unpleasant place to be. Flying insects swarmed, and on the ground were huge cockroaches, millipedes, amphibians (frogs, newts and salamanders) and the first true tetrapods. Amniotes made their appearance in the late Carboniferous. There were reptiles, turtles and lizards. They laid the first eggs that were made up of membranes that protected the egg and a yolk sac that provided food for the growing embryo.

A walk through the swamps of the Carboniferous would not have been pleasant. You would be sweating, swatting at insects and very careful where you stepped or you might inadvertantly become a member of the food chain. The coal mines from the Carboniferous Period caused the beginning of the Industrial Revolution and the pollution of the Earth's atmosphere. It also raised the standard of living for untold thousands of humans. Coal, under great pressure, turns to diamonds, thus the diamond mines in Africa.

Coal, inspite of the warmth and energy it provided, polluted and still pollutes the air with not only carbon dioxide but also sulphur. Coal is the largest source of energy for the generation of electricity. More carbon dioxide is generated from burning coal than from burning oil and twice as much as from natural gas. So, was finding the coal mine a good thing or a bad thing? It was certainly a good thing in that the standard of living for many humans increased by leaps and bounds. Now, however, we are dealing with the downside of air pollution-lung disease, extinction of some species and the big bug-a boo, global warming with all its many effects on weather and the ocean. Remember the ocean covers 71% of the Earth's surface and it could cause a calamity rapidly by rising sea levels and misdirected currents.

The Carboniferous Ocean was warm until the latter part of the period. The ocean was teeming with life. Great plains of crinoids (the first echinoderm) covered the ocean floor. There were corals, bryozoa, brachiopods, ammonites and foraminifera galore. The one-cell foraminifera grew to giant proportions. Normally a foram is microscopic, but in the Carboniferous it grew to huge proportions, several centimeters across, even though it was still one cell and it was given a new name , fusulina.

Foraminifera

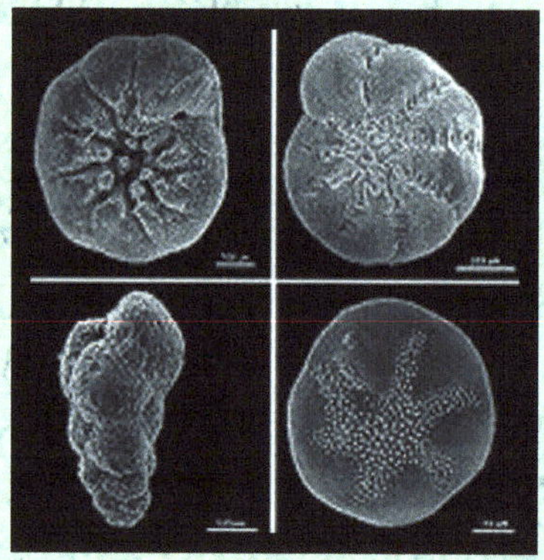

The microscopic radiolarians grew in great numbers and there was evolution amongst the arthropods and the fishes. Sharks continued to evolve. Shark reproduction is interesting in that the male has two claspers (penis-like structures) lying horizantally underneath his body so that he can approach the uncooperative female from either side to send in sperm. The female shark has many methods of feeding the developing embryo. She may lay an egg casing and attach it to a rock or something sturdy. Another method is to let the embryos grow inside her (she has no uterus) and for food they eat each other until one is left to be born.

Toward the end of the Carboniferous, the climate changed abruptly becoming cooler and drier, and caused an extinction of rain forests otherwise known as the Carboniferous Rainforest Collapse. Amphibians fared poorly as did other cold blooded animals. There were, as yet, no warm blooded animals.

The Permian Period

Following the Carboniferous, the Permian Period extended from 299 million years ago to 250 million years ago. It is the the last period of the Paleozoic Era. Its geography consisted of the supercontinent Pangaea. Other land masses merged during this time to create the supercontinent which has hints to where final future separations will take place. Land masses had scattered around the globe, but in the Permian, they all came together to form the great land mass Pangaea. The center of Pangaea lay across the equator. Heat and low humidity blew across Pangaea causing a great desert. It was quite a contrast from the swamps of the Carboniferous. The great Pangaea desert was home to an occasional scorpion, but little else. The desert was also caused by evaporation of the sea, once there, but edged out by land masses. When the water had evaporated it left sand, salt as table salt, and other salts such as gypsum, calcium sulfate.

In the Permian we say goodbye to the charming trilobites. They had endured for 250 million years and specialized into many species, but none survived the Permian. Trilobite fossils are ubiquitous and may be purchased in shops specializing in crystals, geodes and fossils. Also available for purchase are ammonite fossils. Ammonites were common inhabitants of the Paleozoic Seas. Masses of crinoids had begun disappearing in this period, but luckily they persist but are difficult to find. Bryozoans and brachiopods were common but badly affected by the extinction at the end of the Permian. Bivalves such as clams increased and persisted throughout this period.

The dry Permian climate partially decimated the coal swamps of the Carboniferous as well as most of its plants. The climate of Pangaea varied from the end of the ice age of the Carboniferous to warm and dry interiors alternating with cool cycles.

Terrestrial life during the Permian increased and evolved. Plants advanced to seed (instead of spore) ferns. Conifers appeared as well as ginkos and cycads. Conifers produce cones which bear seeds and cycads have seed bearing leaves similar to those now called cycas. *Glossopteris* trees grew in the Southern Hemisphere. They reproduced by seed and grew to 26 feet and tolerated cold weather.

Glossopteris Tree

Gingko trees have persisted through the ages and are much prized today for their fan shaped leaves and beautiful golden fall color.

Insects of the Permian were primarily related to the cockroaches of the Carboniferous. Dragonflies grew in size to a wind spread of 71 cm. or about 4 feet and some believe they were large enough to capture and devour small vertebrates.

In the ocean spiny sharks evolved having paired spines on their pectoral and pelvic fins. They had no teeth and lived by filter feeding, possibly an ancestor of the extant whale shark, the largest fish in our oceans today and also a filter feeder. Sharks with teeth and claspers evolved with spines as well.

By far, the greatest change in the Permian was the large numbers of land vertebrates, tetrapods that evolved. Varanops, large reptile predators were voracious and agile. They were widely distributed as fossils from the USA to Russia and across the southern hemisphere. *Dimetrodon* was the most fearsome. It lived in the early Permian, one of the largest carniferous animals of the time. They were powerful and deadly.

Varanops

Other lizard-type reptiles roamed the land. One with pillar like legs and scutes (raised spikes) on its body must have been a scary sight. It was, however, a 6 ½ foot large herbivore. Another herbivore of the Permian was Moschops. It had a very thick skull and curled horns as in the bighorn sheep of today. The

Dicynodont fossils are so abundant that they can be used for dating rocks. They were more of a pig type animal, but had a middle ear bone it shares with mammals.

So, the landscape picture of the Permian has changed markedly from the preceding period. Now we have a desert central and on either side trees, some familiar, large insects to avoid and fearsome cold-blooded treacherous animals to avoid. It was a dangerous period and one that ended (maybe happily so for herbivores) in a huge mass extinction. Approximately 90% of oceanic and terrestrial life went extinct. The exact cause of this extinction is unknown, but there are several possibilities. Stress caused by cooling, basalt eruptions, loss of sea oxygen, hydrogen sulfide gas permeation, a bolide (meteor) event or maybe a combination of all the above.

The Permian Extinction is important to us today because it caused the death of massive numbers of creatures. So many that as their bodies disintegrated and liquified, a vast amount of oil and gas accumulated. We use these products today, polluting the air with carbon dioxide and sulfur and possibly causing global warming.

Interval

Before going forward into the Mesozoic era in which we will be primarily delving into land animal evolution, a review and more information of the oceanic animals is in order.

There is little known about the ocean currents of early Earth. There were many somewhat small landmasses, thus probably mixed currents. Oceanography has been studied and recorded for the past hundred or more years. There are two major currents on Earth, the upper and the deep currents which are contiguous. When ocean water nears Greenland in the North Atlantic, it becomes colder and denser. This causes it to sink and become the deep current. It passes south to the Southern Ocean and rises near the Bering Strait in the Pacific to become the warm shallow current and continues back across Australia, the horn of Africa and upward in the Atlantic where it becomes denser and colder and sinks again near Greenland. There is also a sink point in the Weddell Sea (near Antartica) which joins the deep circulation. There are great anticyclonic gyres circling the continents. In the Northern Hemisphere the gyre circles clockwise, whereas in the Southern Hemisphere the gyres circle counter clockwise.

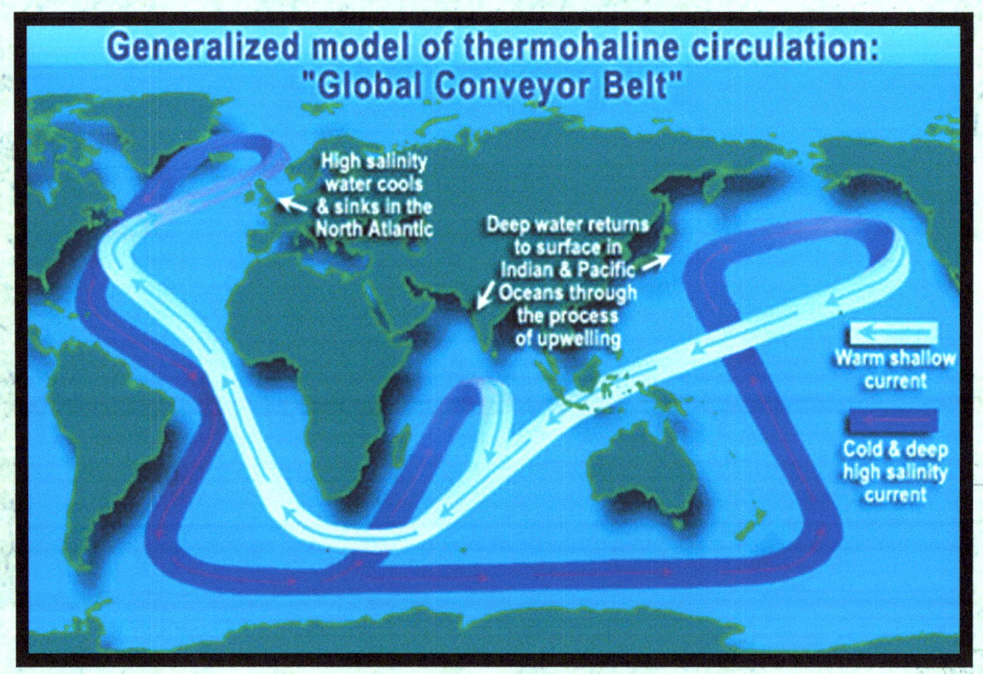

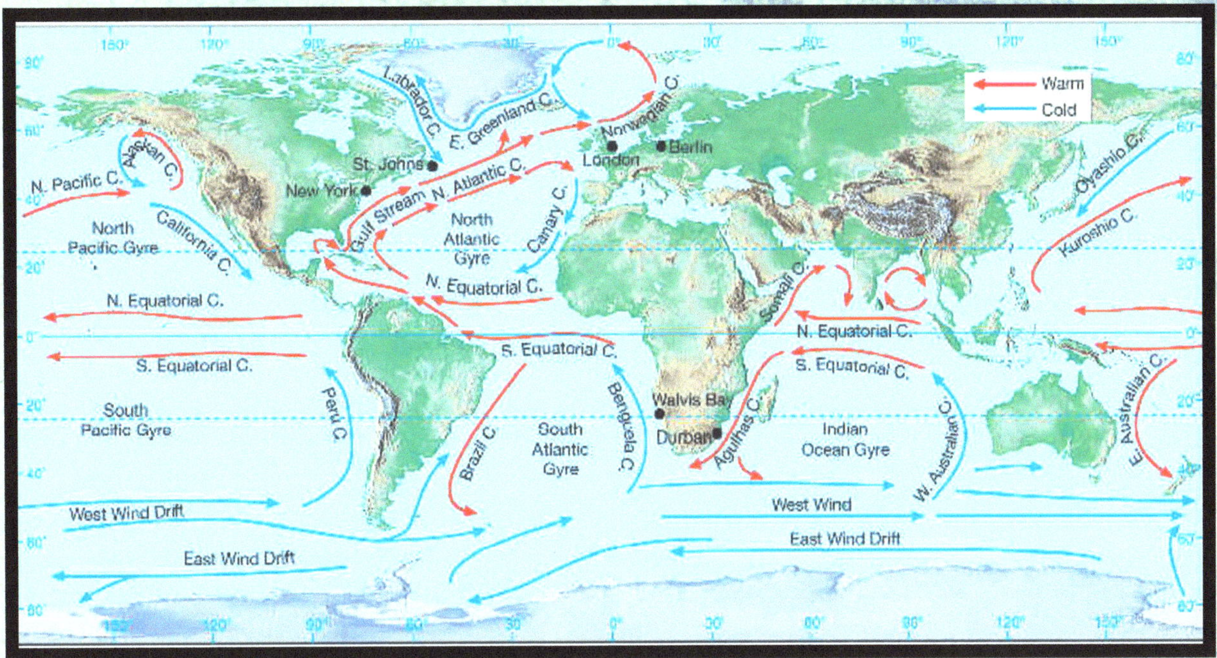

The deep current is important for the food chain in the ocean because it has currents of upwelling near shores which bring nutrients from the sea floor to the upper ocean where plankton are formed-as food for fish and filter feeders. No upwelling-> no fisheries and no food for the web. "How did the nutrients get to the ocean bottom?", you ask. Good question! Creatures that live in the ocean do not foul it with their feces. Their feces is packaged in a polymer wrapping so it drops to the bottom. Over time it breaks down into elements such as calcium, nitrogen, sulfur, magnesium and more. The upwelling takes these nutrients to the plankton, so they can absorb, grow and mature and life goes on.

The fish we catch and eat now are much different from the Devonian (first) fish. Initially fish had hard bony, some calcified, scales covering their body. They gradually evolved into *Telostei* or the slimy small scaled fish we now know. More than half of all living vertebrates at one time were fishes. This may not be true at the present because we have overfished and needleesly killed many fish. There have been loss of habitat (Salmon), pollution and oil spills. Recently the Gulf of Mexico was polluted with thousands of gallons of crude oil killing an unknown number of fish, invertebrates, sea birds and turtles. Fortunately, the Gulf of Mexico has millions of microbes that quickly gobbled up the spilt oil and the Gulf was re-opened for shrimping and fishing three months after the spill.

The Oceanic invertebrates are divided into phyla. All phyla were represented in the Cambrian-some became extinct, others evolved and are with us now. The first phylum includes one-cell animals such as chonoflagellates. The next phylum is sponges which are made up of many chonoflagellates. Sponges were present in the Cambrian. They have remarkable powers of regeneration (if torn apart by currents), they reproduce by budding and there are hermaphrodites and dioecious (sexes separate) types. No matter how they do it, they have evolved into the beautiful, many colored sponges today. Nice to bathe with, too. Their spicules have been discussed under the Precambrian.

The jellyfish phylum is quite large and varied. All but the comb jellies have nematocysts-stingers. Sea anemones were common in the Palezoic, also stony corals or reef building corals. Reproduction is varied, some going into the medusa free swimming type and others remaining fixed as the stony coral. Small fossils of reefs come from the Cambrian.

A word about sexuality is needed. There are two types of hermaphrodites: Simultaneous (the animal has both male and female gonads) and sequential (the animal can change from male to female, protandric or female to male, progynic). Many worms and mollusks are hermaphodites. Some fish that lived in schools (wrass) are progynic hermaphrodites. If the head male of the school is lost (as prey or otherwise), the largest female changes to male. An example of protandric in a mollusk is the slipper shell, *crepidula.* They live in a stack. The lowest one is always a female and the one on top of the stack is male. The sperm trickle down to the female on the bottom. As the top male nears the bottom of the stack, it turns to female.

The simultaneous hermaphrodites are seen primarily in various species of worms. They do not have to look far for a partner. One provides sperm and the other, ova. They never self-fertilize. Another type of reproduction is parthenogenesis. An example is brine shrimp. A male and female produce a female shrimp, she will (all by herself) produce female after female. This is cloning. Sea horses have separate sexes, but the male takes the fertilized ova into a pouch on his abdomen where the embryos develop into perfect tiny sea horses. When they are ready to be born, the male has a type of "labor" in which he pumps the babies out by muscular contraction.

The phylum of Mollusks are varied. They go from a flat shell to bivalves (clams, oysters and scallops), snails with beautiful conical shells (whelk, helmet and conch), nautilus, cuttlefish, squid and the intelligent octopus (thought to be as smart as a dog). The octopus is an escape artist. It can squeeze through a very tiny space. Also, in an aquarium, the octopus is a neatness freak. They will gather up crab scraps from a recent meal into a perfect pile ready for the aquarist to pick up. The cuttlefish is also intelligent and interesting in that their skin contains and is made up of thousands of chromataphores. Cuttlefish can change the color of their skin to blend in with their surroundings, a protective device, and to show alarm or use as an attracting glow when mating. Cuttlefish also have bioluminescense which with added color from the chromataphores causes a colorful, glimmering display. Phylum *Annelida* is a fascinating study (of all things) worms! Remember the Christmas tree worm and the many polychaetes. They are too extensive for this synopsis to cover.

Arthropods are *chitinized* (have a shell) and have jointed legs-think crabs-all kind of crabs. The fossil record goes back to the Cambrian. During their evolution, they were divided into 3 main groups; the Branchiara (lobsters and blue crabs), Cirripedia (barnacles) and Copepods (copepods are part of the plankton as previously discussed). This phylum is characterized by jointed body and chitinous exoskeleton. As the creature grows it must shed its exoskeleton and secrete a new larger one. While they are soft, they are vulnerable as prey. A male blue crab will follow a female in order to be present when she sheds her exoskeleton. Then she is soft and mating can take place. He stays to guard her until her new exoskeleton hardens. Soft blue crabs are reknown for their succulence and are called soft shell crabs. They are deep fried and with a little slaw and ketchup, a very good meal, legs and all. Barnacles are arthropods. The adults cement their rear to the underside of boats causing an increase in drag and the boat has to be hauled out and scraped periodically. Now there is a special paint available that repels the barnacle. If a plank or any flotsam is available at sea, it will be dotted with barnacles. The sexes are separate, so if a male is 3 barnacles away from a female, he must have a very long penis to reach her for sperm transfer-and he does!

The trilobite was an arthropod. Isopods, krill, brine shrimp, all shrimp, sea lice, amphipods and *Limulus*, the horseshoe crab are all arthropods. The phylum, Echinoderm, includes sea stars, sea urchins, sand dollars, brittle stars and sea cucumbers. They have in common a water vascular system and feet called podia which work like suction cups. The sea cucumber has mutable connective tissue. It can be soft and allow the animal to stretch ino a string and later reconnect, or it can be hard and stiff. The sea star opens a bivalve with its podia. It pulls on either side until there is an opening large enough through which it inserts its stomach, digests the body of the bivalve and withdraws the stomach, sated. Most urchins have spines on their exterior, some quite sharp and poisonous and some dull (pencil urchin). The brittle star is known for its rapidity of motion. It can run. The first echinoderm was *Crinoid* or sea lily. It was common in the Paleozoic Seas and it has been discussed previously. There are still some Crinoids but few and difficult to find.

The last of the phyla include the pre-chordates and chordates which had at sometime in their body a notochord. This has been discussed previously as a step toward vertebrates-fish, tetrapods, humans.

Animals with a vascular blood system can be either cold blooded (poikilothermis) or warm blooded (homeotherms or endotherms). Cold blooded animals cannot regulate their body temperature. They rely on the sun. They will hibernate at night and move with the sun as its rises. It will feed only after its internal body temperature reaches a point that allows easy movement. They have a low metabolic rate and require much less food for health. Homeotherms, mammals and birds, on the other hand, have to keep their body temperature at a critical level. This requires eating more frequently (stoking those fires) and having a covering in cold weather-feathers, fur or for us humans, clothing. The demand for food for humans has resulted in increases in agriculture to the detriment of trees and pastures for cows, pigs and other animals we eat. The planet would be better off if all were poikilotherms. All fish are cold blooded, but the blue fin tuna can raise its temperature 2-4 degrees above the temperature of the water when it is swimming fast. It has a special artery-vein complex that creates more heat for faster swimming.

All life, plant and animals on Earth are here because of Evolution. The “theory” is now a fact and can be proven. Darwin knew this to be true but was unable to prove it, but now with knowledge of DNA and genes it can be proved. Natural selection can also be proved. A talk was made to two classes at a middle school by the author. Plankton and other small denizens of the ocean were shown. Afterwards the teachers were asked if they were teaching evolution. Answer: “we are not allowed and encourage the parents to teach it at home.” It came as quite a shock! I like Richard Dawkins’ explanation of evolution. Start with a female rabbit and her mother and grandmother in a row. The row continues back in time and you notice a change in the rabbit until finally it is a shrew, the ancestor of the rabbit. Then make a hairpin turn and watch the shrew for many years, daughter after daughter. The shrew-like animal changes imperceptibly at first then without noticing any big change you find a leopard in its place. What could be clearer? Evolution is very slow. Paleontologists speak in millions of years at a time. You could say God did this, but there is no garden of Eden, no magic flaming bush, only slow change in the gene pool that causes change in the animal. And this change is called evolution by natural selection. Darwin understood that evolution took place by means of natural selection. Humans perform artificial selection on dogs to produce a winner at a show, cats the same, horses for fast running at the racecourse-big money involved here (greed vs. pride?) but in nature, natural selection takes place over much more time. As the gene pool changes, the individual with genes (say for fur) for a thick fur growth in the arctic would pass his/hers on to the next generation. If the gene for thick fur was not passed along due to the gene mutation or lack thereof, the individual would die and his/her gene would not be passed along. Or more simply put, natural selection is the process whereby organisms better suited for their environment tend to survive and produce more offspring. Even the Pope believes evolution is true.

You should be at least somewhat familiar with the term "Gaia". Many scientists accept this hypothesis of the Earth's existence. Gaia is a term for the Earth as a self-regulating system made up of all organisms, land rocks, oceans and the atmosphere closely adherent as in one system which regulates itself to be favorable for life. There is a conflict among scientists concerning the Gaia hypothesis as opposed to Neo Darwinism which includes genetics.

In closing this interval, it's important to remember that all dates given for eras and Periods are not etched in stone. (Pardon the pun) The study of paleontology is constantly finding previously unknown fossil fauna. In addition, vertebrates are found even in the Cambrian and did not require a special step in evolution.

While studying the land animals, those in the ocean continued to evolve and grow in great numbers in order to recover from the Permian extinction.

The Triassic Period

The Triassic Period is the first period in the Mesozoic Era. At the end of the Permian it is estimated that 90% of all species went extinct. In addition Pangaea was beginning to break up so that some species were separated from their kin and could, therefore, evolve differently. Australia separated from Pangaea early and became an island. Those animals evolved into marsupials, large and small found only in Australia. The opossum found in North America is also a marsupial. It can also "play dead". If frightened and fearing for its life, the Opossum will lie on its back, the eyes are open but glazed white and there is no evidence of respiration. It truly looks dead, but after the danger is past, it quickly gets up and runs away or climbs a tree. (author's experience)

The Traissic extended from 250 to 200 million years ago plus or minus a few million years. It was a period of transition from the Paleozoic to Mesozoic. The first flying vertebrates, pterosaurs, evolved during the Triassic. At the beginning of the Mesozoic, Africa was still joined with Pangaea and shared its fauna, mainly theopods, prosauropods and other carnivorous reptiles. The list of vertebrates that evolved during the Triassic is huge as was their size. All were tetrapods, cold blooded reptiles, both carnivorous and herbivorus. Some species lived in the shore line water and used their limbs for swimming and fed on fish. Many of the terrestrial animals were fearsome and huge. They were not yet dinosaurs, but headed in that direction. Raurisuchia was a crocodile-like archasaur. They were giant quadripedal hunter omnivores with rows of protective plates on their backs. One such is the Postosuchus.

Postosuchus

There are many crocodile-like Parasuchus which grew to sizes as large as *Tyrannosaurus Rex* would be. It was primarily a fish eater but found world wide.

Parasuchus

A primitive *archosaurs* was the Euparkeria. It was a reptile that lived in what is now South Africa. It was a relatively small carnivore and had a mouth full of razor sharp teeth. All the better to eat you with! The Karoo Desert of South Africa contains foot prints of this animal. In Argentina, the fossil of Herrerasaurus was found and is thought to be a primitve dinosaur. It walked on its back two legs, had sharp claws and serrated teeth. It could run fast to overtake its prey.

Herrerasaurus

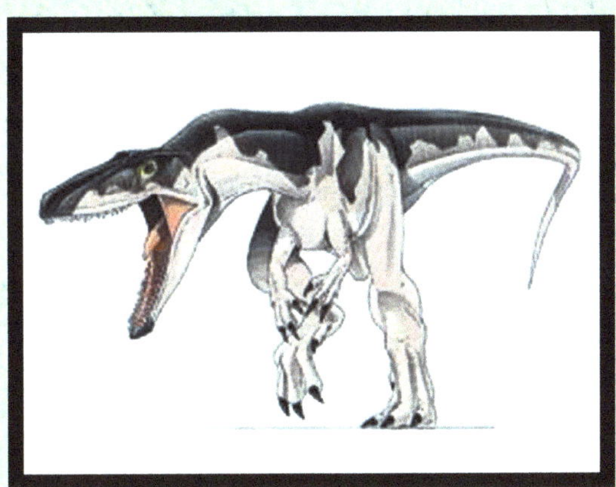

The Eudimorphodon, a pterosaur, was a flying reptile, with fossils found in Northern Italy. It was capable of flying as well as gliding. The Plateosaurus of the late Triassic fossils found in Norway, Switzerland, Greenland and Germany had a long neck for tree browsing. It was large-33 feet long and weighed more than a ton. Its teeth were rough bumps typical of a vegetarian yet you would hardly want it for a pet. There are many more of these amphibians and reptilian creatures' fossils found and identified. Should you wish to see more, please see *Prehistoric Life* in the bibliography.

Plateosaurus

In the ocean, the Ammonites, crinoids, gastropods, nautiloids, bivavles and brachiopods were present, having survived the previous extinction. The plankton became more diversified and the Triassic oyster evolved. All the bivalves were more numerous during this period. The first mammal probably evolved during the Triassic via the Cynodont.

Cynodont

The earliest turtles and some small lizards appeared and it is thought that there were mammal-like successors to the Therapsids, having fur covering the skin and higher metabolic rates. Therapsids are reptiles and reptiles are amniotes meaning their embryos are covered with an amniotic membrane and fluid. They are considered to be pre-mammalian.

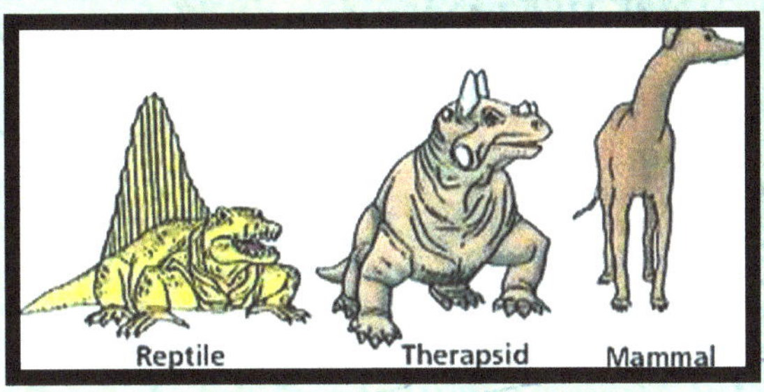

In Australia there are 2 animals that lay eggs. After the egg opens, the embryo ambles somehow to the chest of the mother where milk is secreted from her sweat glands. They were the duck-billed platypus and the Ecidna. They are extant and thought to be at least one of the connecting links from egg layers to mammals. The cynodonts are a large group that may include true mammals. The first cynodont arrived in the Permian and many of them prospered in the Triassic. They developed hair, whiskers and canine type teeth which suggest they were warm blooded. The next extinction which wiped out the dinosaurs may have given these creatures a chance to evolve.

The Triassic enjoyed warm temperatures and much of the continent was dry. This may have been one of the hottest times in the history of the Earth. The Triassic brought many new plants to add to the flora. There were new ferns, conifers (such as the Norfolk Island Pine) and new lycophytes such as the Pleuromeia which was 6 feet high and distributed world wide.

Pleuromeia

The Decrodium trees reached 98 feet in height and could have been browsed by the herbivores. Cyads increased in number and had a cone shape. Ginkgo continued to grow. It is a flowerless seed plant kin to gymnosperms. Its living relative is the lovely maidenhair tree. Conifers grew 30-40 feet tall.

The Permo-Triassic extinction dealt a huge blow to insects, but by the late Triassic both insects and the pollen they fed on had recovered. New Triassic insects were bees, ants, wasps, mosquitos, gnats and sawflies. All of these are extant insects and their body plans are unchanged.

An extinction of unknown cause occurred at the end of the Triassic. Many of the niches previously held by reptiles could be filled by other and more fearsome animals, the dinosaurs.

You would be foolish to even want to visit Earth during the Triassic. It was noisy with grunts and squeals from the various reptiles wanting to gobble you up.

The Jurassic Period

The Jurassic Period extended from about 200 million years ago to about 145 million years ago, or from the end of the Triassic to the beginning of the Cretaceous Period. It is in the middle of the Mesozoic Era or the "Age of the Reptiles". The term "Jurassic" comes from the Swiss Jura Mountains. The word "jura" means forest mountains.

The supercontinent Pangaea was in the process of beginning to break up into Laurasia, the upper continent and Gonawanda, the more southerly continent. The Atlantic Ocean was just beginning to open up so that South America began to rift away from Africa and a hint of North America and Europe would come from Laurasia. As yet, there was no land at either pole and no icecaps. The climate was warm and humid and there were temperate rain forests in both hemispheres. The oceans had a chemistry change resulting in carbonate hard grounds (ocean floors) and an increase in invertebrates with calcium carbonate shells. Coralline algae appeared for the first time. On land there was an increase in batholiths, intermediate hard rock such as granite and quartz. In the oceans more corals, ammonites, crustaceans and Brachiopods flourished and lobsters increased as well as early squids. The Hybodus shark was more robust than modern sharks, but their teeth were similar in that there was continous production. Amphibians like frogs (jumping frogs), the snake-like Eocaecilia with 50 vertebrae and the fish, Lepidotes, with thick scales and rows of peg-like teeth swam throughout the ocean.

Lepidotes

Teeth

In addition the oldest turtle appeared as well as the Jurassic dolphin. Icthyosaurus joined in and could swim an estimated 60 miles an hour. Mary Anning helped supplement her family's income by walking the beach on England's south coast, collecting fossils which they sold. She was the first to spot an Icthyosaur on a high bank. Imagine the thrill she must have felt. She, although a woman, was inducted as an honorary member into the Dorset County Museum. Montrous crocodiles evolved which could eat their weight in fish and squid daily. A small Pterosaur, Dimorphodon, could fly and dive for fish. He was clumsy on land and had a huge puffin-like beak. The long tailed Pterosaurus was a better flier with long tail and narrow beak. The Pterodactylus had claws on its wings and a roseate crest on its head.

Pterodactylus

So where do the reptiles end and the Dinosaurs begin? In the latter third of the Jurassic Period , small mostly herbivourous therapods evolved. The name Theropod and dinosaur are used interchangably throughout most of the works of Paleontology. They are still reptiles, i.e. cold blooded but dinosaurs are more fierce, they have shorter forelimbs and probably fed on "lower" reptiles. Some lived in or near oceans and fed on fish. Many large dinosaurs had bumps or ridges on top of their heads. Some had horns such as Allosaurus.

Allosaurus

Many more fossils were found around the world. Most from the late Jurassic were "lizard-hipped" dinosaurs referring to the shape and structure of the hip.

The Pterosaurs ruled the skies during the Jurassic and eventually their niche was taken over by birds. Some dinosaurs had normal length forelegs; some had extremely long necks like the Brachiosaurus, which was most likely a herbivore and browsed at tree tops. Its neck measured 75 feet.

Brachiosaurus

The Stegosaurus was called a "plated lizard" for its bizarre diamond shaped plates along its back.

Stegosaurus

The plants of the Jurassic have been studied and dispersed parts preserved in Yorkshire, England. A more humid atmosphere than the Triassic allowed lush jungles to grow and cover much of the land. The ferns continued to grow along with the conifers. Gingkos reached their maximum diversity during the Jurassic. Horsetails and cyads were common. Besides Horsetails, *Selaginella,* grew in the Jurassic and is extant. It is mostly found on tree branches now. From the group *Bennettitaleans* came the Williamsonia, which grew to 7-8 feet and had a thick trunk. At the top, it had a thick dome from which came long fern-like leaves and possibly flowers. This may have been the first flowering plant. The flowers were named *Weltrichia.* They were star shaped and pollen bearing, but they have never been found attached to a tree.

Williamsonia

Michael Crichton wrote the book and Steven Spielberg made the movie, " Jurassic Park". It was a smash hit and put dinosaurs to the public front like nothing before. There was absolute dinomania. From almost complete obscurity to a splendid take over, they rose in popularity overnight. Every child had a miniature dinosaur toy, a T-shirt, mug, etc. *T-Rex* was featured at all museums. Stephen Jay Gould, reknown Paleontologist, had a chat with Mr. Crichton. It seems only two of the dinosaurs in the film were from the Jurassic. All the rest were from the following Period, the Cretacaous. Even the popular *Tyrannosaurus Rex* as well as the Velociraptor were in the late Cretaceous. But how would "Cretaceous Park" sound? Not so hot, huh? So "Jurassic Park" it was, allowing for artistic license. Actually, Mr. Crichton did not know from which Period the dinosaurs came. He said "I went with what looked best." Spielberg used all his technique and imagination in depicting the bodies (computer generated) and expressions of the dinosaurs. He even put an evil glint in the eye of *T-rex.* Dinosaurs reached the height of popularity in the 90's and never had any competition. Millions of dollars were made during the frenzy. There are also Jurassic Park two, three and a fourth in the works.

The Cretaceous Period

This geologic period was dated from 145 million years ago to 65 million years ago. This is truly the Age of the Dinosaurs. They ruled the Earth. Turtles also evolved during this time period and likely developed its hard shell covering for protection. Dinosaurs lived for 160 million years. There were about 9,000 species and their fossils have been found on every continent on Earth. They were herbivorous, omnivorous and carnivorous feeders. They had skin with thick large scales, some with horns, crests, plates and other display structures. Dinosaurs were cold-blooded egg layers. Some had shortened forelegs and large tails thought to be for balance. Most dinosaurs were 8-10 feet tall, but many others were huge, topping off at 39 feet for *Tyrannosaurus Rex.*

Tyrannosaurus Rex

The omnivore and carnivore dinosaurs were aggressive and cunning regardless of size. They could weigh up to 70 tons. Some dinosaurs lived in the sea and ate fish, but they were not equipped with gills or webbed feet. These could move fast to catch fish prey using their extremities.

Dinosaurs had two types of hips: the sprawling frog-like hip or the erect hip such as humans have. They probably evolved during the Triassic Period.

Dinosaurs (including birds) were much like our modern crocodiles. Birds evolved from frog-hipped dinosaurs. They developed hairy-feathers, then true feathers and became warm blooded.

The largest dinosaurs were the Brachiosaurus, about 85 feet long and weighing up to 50 tons, and the smallest about the size of a crow. Some dinosaur fossils were found in herds-many creatures together for protection or on a hunt. One fossil found in the Gobi desert turned out to be two different species (a Velociraptor and a Protoceratops) engaged in a fight.

Dinosaurs probably did not roar as depicted in the film. No sign of vocal cords was found at dissection where soft body parts were still present. Even birds do not have laryngeal vocal cords. They have a structure called a syrinx at the bifurcation of the trachea into bronchi.

A 68 million year old *T.rex* was found in Montana recently. It had some fossilized soft tissue. This is an ongoing investigation.

It was not until 1950 that the extinction event, at the end of the Cretaceous, drew scientific attention. Many theories were set forth to explain the extinction, especially of the dinosaurs. It was only in 1979 that the presence of iridium (found in meteorites) that further investigation was warranted. The blast had caused a crater 10-13 miles across and blown a cloud of dust and debris into the upper atmosphere, large enough to darken the planet for a year or more. The impact site in Mexico on the Yucatan Peninsula was found in 1991. It caused global refrigeration. The only survivors of this terrible cataclysm were carnivores who ate carrion, birds and small mammals. Crocodiles survived by eating detritus or by hibernation.

During the Cretaceous Period the continents were on the move. The North and South Atlantic Oceans were formed, Australia had broken away from Pangaea and North America and Eurasia were forming. Huge seas covered many of the present day continents. The climate was variable. About every two million years there were changes in temperatures from high to low. Anarctica began forming ice sheets, but by 70 million years ago, the Earth warmed. Even high latitudes had temperatures of 80 degrees and the oceans as high as 97 degrees Fahrenheit.

Aside from dinosaurs, the Cretaceous Period is also known as the period during which the first flowering plants, angiosperms, appeared. The structure inside the flower is sexual i.e., ovary (female obviously) and the stamen with its anther and pollen, the male parts. Because pollen from the anthers had to be carried to another flower's stigma to ovary, it has been thought that there was mutual coevolution of flowering plants and insects. It is a neat hypothesis, however it is possibly not true. Insect fossils have been found since the Carboniferous, went almost extinct during the Permian extinction, but began again during the Triassic Period and continues to the present. Bees had been pollenating gymnosperms (cone bearing trees) 100 million years before angiosperms evolved. It would seem logical that insects and flowers co-evolved.

There has been some difference of opinion among paleontologists regarding plant-insect co-evolution. Some believed insects were old enough to transfer gymnosperm pollen 100 million years before the Cretaceous. It seems that there may have been an explosion of other pollen carriers (butterflies, wasps, more species of bees, moths, crickets, lacewings and more) occurring as angiosperms evolved. This would be termed co-evolution. Showy flowers evolved during the late Cretaceous such as magnolias, witch hazel and other flowering trees. Leaves fossilize much better than flowers, which tend to fall apart, consequently the fossil record for most small plant flowers is poor. Fossil plants or leaves are useful in plotting the many climate shifts that occurred as plant life evolved.

Orchids have been mentioned before as the Darwin Orchid with its long spur. Orchids are the most highly evolved of all plants. There are terrestrial and epiphytic types and can be found at almost all latitudes. Terrestrial orchids grow with their roots in the ground but must have a certain fungus nearby in order to thrive. Sadly many of these plants are taken and lost to the taker as well as the world. Epiphytic orchids grow on branches of trees mostly in warm latitudes. Their nutrition comes from rain laced with bird feces. There are many species some of which have evolved devious methods to attract insects for fertilization. There is the scent attraction, most alluring. The whole of the Brassia resembles a spider urging a real spider into its "web". Then there are orchids whose reproductive structures grow to look like a female insect. The male insect then tries to copulate, picking up the orchid's pollenia in the process and eventually transferring it to another orchid of the same species, again a false copulation occurs. There are also orchids whose sexual parts resemble a stinger insect, stinging takes place and again the pollenia are transferred. Orchid seeds develop inside a sac. When they are mature the sac bursts with the wind and thousands of very tiny seeds are wafted away. Probably only one or two will land in a place propitious to its growth. Other highly evolved plants are the carnivorous plants, plants equipped with enzymes to digest insects which they attract in various ways. These are the Venus fly traps, Sun Dew plant, Pitcher plant and many more. Insects provide these plants with growth, but their energy comes from the sun. Termites are the first insect to live in colonies followed by ants and bees. This is the first of the colonies with a division of labor. They typically have a fertile "queen", fertile males and infertile workers.Oceanic invertebrates and vertebrates continued to grow and evolve new species. The cephalopods, fishes and sharks appeared and the gar pike appeared just as it is today. Frightening aqueous dinosaurs such as the Elasmosaurus had a very long neck and probably ambushed its prey (fish mainly) by manuevering its neck. In the late Cretaceous the Pteranodon, a pterosaur (flying dinosaur) flew across the seas living on fish. It was about 6 feet long and could glide much like an albatross. The Cretaceous also saw the fearless crocodile evolve. It was 39 feet long, weighing up to 10 tons. It is thought to be the precurser to the modern day crocodiles.

The Cretaceous mammals were small rodent-like creatures covered with furry hair. One such is the Eomaia.

Eomaia

They had a placenta, gave birth to their young and nursed them from mammaries with milk. The non-placental animals are the duck billed Platypus, Echidna and marsupials found in Australia which you recall has just moved away to form its own island, leaving marsupials to diversify, as they did. The predecessor of humans, however, had a placenta and mammaries and was a warm blooded, fur covered-rat! These mammals were not totally exterminated in the late Cretaceous extinction and together with some insects and Pterosaurs, which evolved into birds, lived to re-establish life on Earth.

The Tertiary Period

The Eocene Epoch and The Miocene Epoch

We have concluded the Mesozoic Era and now will enter the Cenozoic Era, the early part of which is the Tertiary Period. The entire Cenozoic Era is also divided into Epochs. Refer to your Chronological Chart for a more clear understanding. We will work through each Epoch giving the time span and the name of the Epoch to which you can refer if necessary.

Imagine in your mind's eye how Earth looked after the smoke and dust had settled and the Sun once more could be seen. There were vast empty plains yet to be colonized by new life. Destruction of trees, plants and animals must have left a queer odor and a stark scene. It was about 65 million years ago and there were a host of evolutionary niches to be filled. After a questionable period of time, life returned to Earth, but in a very different direction. In the Paleocene, 65-55 million years ago there was a period of darkness and cessation of photosynthesis. The temperature plummeted. All dinosaurs (save the flying ones), most large reptiles, plants and trees were decimated. The early part of the Paleocene was a period of transition. Dust settled, the sun rays gradually returned and the temperature gradually warmed, including the oceanic temperatures. The warmer water led to a return of life in the ocean i.e., corals and many other invertebrates. Grasses and seeds began to grow putting new life on the planet. During the Paleogene which includes the Paleocene, the Eocene and the Oligocene (65 to 23 million years ago), the Earth's continents moved to roughly the position they are at the present. Both poles became icy. Plant life returned rather quickly consisting of ferns and conifers, grasses and sedges, as well as flowering plants, angiosperms. Legumes appeared and many other plants and trees that we now have. Mammals rapidly evolved capturing all the niches previously owned by dinosaurs. The Diatryma, a flightless bird, from the Eocene, lived in forests of the western United States as well as Europe. They were large 6 foot birds and active hunters, but were no match for the mammalian carnivores.

Diatryma

Dugongs and Manatees descended from the *Ambulcetus*, the Eocene "walking whales", so called because it had four walking legs.

Turtles have connections with both land and water. Sea turtles breath air just as their land kin do. There is also a land connection in that sea turtles return to land (beaches) to lay their eggs.

Duck-like birds were found in Wyoming and one of the first fossil mammals, Leptictes, was found in North America.

Leptictes

The Andrewsarchus, a placental mammal, named for the paleontologist who discovered its fossils in Mongolia, was a formidable predator with sharp teeth.

Andrewsarchus

Bat fossils were found in Wyoming. Anteaters have been dated to the Eocene as well as the Protohippus, a small four-toed horse. In China, marmoset fossils were found and the Leptomeryx was one of the common placental mammals of the Eocene.

Leptomeryx

The Leptomeryx had horns or antlers and is thought to have been a relative of all ruminants such as deer, cattle and camels. The *Eosomias* evolved during the middle Eocene. It is thought to have been the first primate. The fossils were found in China. It had grasping hands and the appearance of a tiny marmoset. This suggested that the birthplace of primates was Asia rather than Africa as previously thought, however the African origin prevailed.

It took about 10 million years for insects to recover from the Cretaceous extinction, but once back, they quickly made up for time loss and now are thought to be the most numerous of any life form, except bacteria.

Many of the birds we know now began life in the Eocene.

Throughout the Eocene there were hoofed mammals that could enter the water to hunt and by the end of the Eocene, they became fully aquatic whales. Fossils of whales have been found that dated back

40 million years and they were much like the whales today. Whale DNA relates them most closely to the hippopotamus. Its evolution proceded to dolphins and otters.

Recently a 36 million year old fossil of a penguin was found in Peru. This puts it neatly in the Eocene. It weighed more than 100 pounds and had reddish colored feathers, quite dissimilar to the modern penguins.

It was not known until fairly recently exactly what had caused the dinosaur extinction. Paleontologists had many theories one of which was a asteroid impaction. But how could one prove this 65 million years later? A geologist, Walter Alvarez, began looking for iridium. It is the only element not found on Earth. Lo and behold, much iridium was found around the collision site in Mexico and geologists mapped the crater 100 miles wide. The case rests.

The dates for the Miocene Epoch are 23-5 million years ago. The Miocene is not marked by specific boundary events. At the end of the Pliocene (and therefore Miocene), we will have completed the Tertiary Period or about one million years ago. Please refer back to your chronological chart for a clearer picture of this time period because many species appear in rapid succession. During the Miocene, mammals and birds became well established. Three toed horses and hornless rhinos evolved. Ducks, owls and crows made an appearance and marine birds reached their highest diversity during this epoch.

Brown algae and kelp appeared to support new species of aquatic life such as otters, new species of fish and invertebrates. Approximately 100 species of apes roamed over the Old World. It is not clear if Old World apes were precursors of hominids. The ungulates such as pigs and hippos were soon to be followed by gazelles and goats. The Bering Strait provided a land connection to North America from Eurasia and the ancestors of modern Bison and Big Horn Sheep arrived in North America. Rodents led the way for mice, voles and hamsters. The land bridge provided a cross for porcupines to colonize Eurasia. About 20 million years ago in the Pliocene, apes spread from Africa to Asia and were the forerunners of modern orangatans.

The largest shark appeared in the Miocene---the Carcharodon.

Carcharodon

Its massive mouth with many rows of serrated teeth preyed on every living animal in the ocean. It was a relative of the modern white shark. The modern cormorant, pelicans and gannets are members of the *Phalocrocorax* genus. Neogene meaning "newborn" is the time period when many of our modern animals evolved as you can ascertain by now. The Neogene includes the Miocene and Pliocene Epochs. On land the climate temperatures were warm to cool and the poles were beginning to ice over. Carbon dioxide levels, as ascertained by drilled cores, were fairly stable during the Neogene. A northern ice sheet was formed over Greenland, forerunner to later glaciation.

The Isthmus of Panama provided a path of movement for many species. It also cut off warm equatorial ocean currents and an Atlantic cooling cycle began.

In the ocean, many species evolved such as the Oliva gastropod, the coral Cladocora of Cyprus, many tube-worms, limpets and others too numerous to list here.

Plant life exploded and the great Himalayas, Rockies and other mountain ranges rose as plate tectonics caused land movement. There were grasslands, forests of most of the trees we are familiar with plus the swamp cypress. Poplars, elms, beech and maples evolved.

Other animals of the Neogene include Pliohippus, a one toed horse which was not the ancestor of the modern horse. A "terrible beast" was a relative of the mastodon. It was Deinotherium, about 14 feet tall and with massive tusks and an elephantine nose. It weighed up to 14 tons.

Deinotherium

Another placental mammal was the Dryopithecus. It ranged in Europe, Asia and Africa and had a chimpanzee-type body and walked on all fours.

Dryopithecus

This is a good time to stop, close your eyes, and try to picture the Earth with its clear blue oceans, lovely green forest, bright flowers here and there, crystal lakes and rivers, birds singing, other animal noises, most non-threatening, animals grazing, no roads, clear pure air-a beautiful blue unsullied planet.

The Quaternary Period

This is the most recent of the three periods of the Cenozoic Era. It dates from 2.6 million years ago to the present. It includes two epochs, the Pleistocene, 2.6 to 0.78 million years ago and the Holocene 0.7 million years ago to the present. The beginning of the Pleistocene marks the beginning of glaciation.

The oceans and continents were greatly influenced by climate change during the Quaternary period. Following every other ice age, there were changing patterns of lakes, rivers and inland seas. There were glaciers to 40 degrees latitude (central United States). Variation of degrees of glaciation and climate in general are based on the fact that there are variations of incoming solar radiation as well as variations in degrees of perigee and apogee, the distance Earth is from the sun.

Earth's oceans and continents were influenced by climate during the Quarternary more than at any other time. During this period, humans who had previously lived as hunter-gatherers began to establish more permanent living quarters and domesticated animals. In North America, the last glacial advance occurred 15,000 to 20,000 years ago. In order to cope with the cold, giant mammals grew thick fur coverings. This is the age of the bison, the wooly mammoth, deer and cave bears. The saber-toothed tiger was muscular and deadly. It disappeared during the last ice age. In Siberia, a frozen baby mammoth was found. It was about 4 months old when it died and carbon dates to 30,000 years ago. Canis dirus, a dire wolf, lived in the Pleistocene as did a giant bear, Arctodus. In South America, rhinoceros fossils were found that resembled those found in Eurasia, a classic example of convergent evolution. South America split from Africa in the Jurassic. Beavers, condors and moas (flightless birds) were present in this period as well as the armadillo. A relative of the Komodo Dragon lived in Australia. Now we find the Dragon only in Indonesia, on the Island Komodo. Even its saliva is poisonous. Horses were common in North America and probably crossed the Bering Sea bridge into Asia. Wild asses, zebras and Przewalski's horse evolved.

During the late Pleistocene and early Holocene, Aurochs, now extinct, were the ancestors of today's cattle, somewhat larger than cattle are now. They developed in India. About 9,000 years ago cattle became important to humans. They could not be tamed but they could be caught and enclosed. Both the meat and the milk became an important part of the diet of early humans.

Madagascar is an island off the coast of Africa. It split away from Laurasia 88 million years ago or in the Mesozoic. It is the 4th largest island in the world. When land split away from a continent very early, its diversity of plants and animals was greatly increased. This is the case in Madagascar. In other words, if part of a group of animals is moved away from the other ones, the resulting two animals may evolve into a separate and distinct species. More specifically, the lemur on Madagascar evolved into a diversified lineage, all endemic to Madagascar, whereas the lemurs left behind in Africa evolved into primates as monkeys. Characteristics of primates include large head and high forehead, eyes that look forward causing stereoscopic vision, mobile arms, two mammaries, opposable thumb and nails instead of claws (perhaps the most important feature). The lemurs and tarsiers date from the Eocene (55 million years ago), however the date may have been earlier and fossils not found. But this is guesswork and we would do well to stick with the Eocene.

Out of the mammalian tetropods, only a few were capable of being domesticated. Most were domesticated in Asia where humans began cultivating crops. The dog was the first at about 15,000 years ago in the Miocene. It was probably a lost wolf pup and when it mated the genetic outcome was part tame and so on through the years until it was completely tame and had been taught to help with hunting, protecting the home by guarding and enjoying petting. About 10,000 years ago, still in the Miocene, sheep, goats, pigs, cows and cats were domesticated. Later on in the Pliocene about 5 million years ago chickens were domesticated and a new food source, eggs and chickens, became available to Homo erectus and later Homo sapiens. Other animals that could be domesticated include llamas, alpacas, horses, dromedary camels, donkeys, water buffalos, bactrian camels, yaks and turkeys. Feral cats are with us yet we believe they may have domesticated themselves, being drawn to enclosures with grain that in turn attracted rodents. So cats just hung around the food source and gradually became pets for humans.

Human Evolution

Primates sprang from Amniotes (egg layers whose eggs have a soft membrane). The duck-billed Platypus as you recall laid eggs but also made milk even though they had no mammary glands. They also have no placenta. Marsupials have no placenta but do have mammaries that make milk. Primates, then, are animals that have a placenta, bear live young, have mammaries and are arborial (monkeys) or live on the forest floor (gorillas). Chimpanzees have all these attributes, plus DNA that is 96% similar to humans. Early humans are called Hominids. In addition to placenta, live newborns and mammaries, humans are bipedal. We walk upright on two feet. Could this shift in body weight and carriage be the etiolgy of human back problems? Backaches of one kind or another are very common. Conditions such as slipped disks (remember the notochord), scoliosis, kyphosis (round back) and sacro-iliac problems and hernias are extremely common. The advantage of being able to carry and use one's hands meticulously, as in surgery and art, surely outweighs the accrued problems stated above. Hominids diverged 5-8 million years ago in Africa and their next closest relatives are the gorillas, orangutans and bonobo. The beginning of Hominids occurred about 7 million years ago in Africa. The early hominids were called *Australopithicus.* The word means southern apeman. There were several species of *Australopithicus,* but the best known is "Lucy", *A. Afarensis,* 3.2 million years old, found at the Leaky camp in Tanzania. Her brain was not as big as current humans, but her body was also small. About 2 million years ago it is thought that tool making occurred. The earliest Homo was Homo habilis, thought to be the first tool maker.

Homo Habilis

His nickname is “handyman”. He had a bigger brain relative to body size. The first tool was a flake of stone made by hitting one stone with another in such a way that a sharp stone flake comes away. Many flakes with sharp edges could be made from one stone. These sharp edges were used for cutting-meat, skin, tendons, plant stalks and stems. Early Homo lived on the plants, nuts and whatever small animals they could kill. One of the most significant anatomical differences between Hominids and humans and other primates is bipedalism and shorter arms. Bipedalism frees the arms for carrying and making tools. Bipedal humans were freer to leave the jungle and go out into the meadow lands. *Homo ergaster,* meaning “workman” was closely associated with tool making. He is considered by some scientists to be the ancestor of all Homo species, including Homo sapiens (humans).

Homo Ergaster

His skeleton was found in Kenya 1.5 million years ago. *Homo ergaster* may have been the first Homo without much body hair. More recently an article in the Smithsonian, March 2010 issue, by Ann Gibbons describes fossils of an ape-like creature that lived 4.4 million years ago. It was found by Tim White. He calls her "Ardi". Her skeleton was found in the Afar Desert of Ethiopia. They also found fossils of humans only 160,000 years old-a virtual storehouse of paleontological finds. "Ardi" is short for *Ardipithecus ramidus.* Then there was the "Java man" later classified as *Homo erectus,* one of our direct ancestors found in Java and 1.8 million years old.

During the evolution of humans, the brain size gradually increased over time.

Ardipithecus ramidus

Homo neanderthalensis fossils were found all across Europe and Asia. They had advanced hunting tools and were artistic. Most scientists believe they went extinct because of the cold climate. The Cro-Magnons lived in France and were tool makers. They were tall and had complex societies, produced art, made ornaments and had language and may have replaced the Neanderthals. Cro-Magnons gave rise to the people of Europe.

More recently, hominid fossils dating to 1.98 million years ago have been found in Malaya, South Africa. Lee Berger, Paleontologist, calls it "the Rosetta Stone to the origin of Homo." It consists of fossilized bones of several Hominids with opposable thumbs and having a height taller than Lucy. Its scientific name is *Australopithecus sediba* and it is thought to be an intermediate form between the primitive australopaths and our genus.

Tools, language, domestication of animals (the first was a dog, probably a lost wolf pup) and control of fire are the essential hallmarks of *Homo sapiens.* There was most likely multi-regionalism in the arising of *Homo sapiens,* but most modern humans evolved in Africa and spread to all continents. It would be worth a trip to Spain, France and Russia to view the cave art and other artifacts left by our ancestors. The cave art, all of animals, is said to be beautiful. Animals, both domesticated and wild, were very important to our ancestors. Their way of life, from hunter-gatherers to communities, agriculture and domestic animals was changed forever.

The "Human Revolution" is thought to have taken place 50,000-40,000 years ago. Cave paintings in France and Germany are dated to about the same age i.e. 32,000-36,000 years ago. Mostly large animals are painted and said to be "breathtaking" in their strength and beauty. There are more than 50 caves painted by early humans. They are located in France, Spain, Africa, Asia and America. Some have been closed to the public due to fungal growth destroying some of the art. They are thought to have been painted about 13,000 years ago. Most of the paintings are of large animals-horses, bison, ibex and other horned animals. They are large, as large as the painter could reach on the cave wall. They are beautifully portrayed, many as if in motion. The animals are outlined in black manganese oxide or charcoal mixed with animal fat. A film has been made by Werner Harzog called "Cave of Forgotten Dreams". You walk through the Chauvet cave in France. This cave is closed to the public in order to save it for the scientists. See the film if you have an opportunity.

Homo sapiens are the only group of animals on Earth to have consciousness of self, to have intelligence enough to learn to fly into space and construct a complex computer system in one generation. We are the only animals that know we will die. We have conquered much disease and our knowledge is worldwide. Most of the animals we now know were domesticated by about 10,000 years ago. The domesticated horse made a clear advantage to humans. They are beautiful, strong and swift of foot. Will Rogers said, "A man that thinks a dog is his best friend, never had a horse." A bit of heaven is to kiss a horse's velvet nose.

Now we are entering the Anthopocene, the age of man, and a new name for a new geologic epoch. It describes the huge impact man has had on this planet. Our population has grown to 7 billion, living primarily in big cities. It is the oil century. Thousands of gallons are pumped up from the Permian Period to propel the millions of automobiles and trucks; thousands of these pumped gallons have inadvertently been spilled into otherwise pure, fresh water or into the oceans. Species are down due to pollution and loss of habitat. Coal from the Carboniferous is mined for production of electricity, spewing carbon dioxide and sulphur into our once pure air. The ozone protection from harsh rays is diminishing. Huge numbers of trees have been cut down to build abodes and make more land for pastures and agriculture. Superbugs are growing in the water in Calcutta. Many antibiotics have become useless as disease bacteria develop more and more resistance. We have dumps overflowing with chemicals and plastic. The oceans are being acidified due to the huge volume of carbon dioxide dissolved.

Nearly three-fourths of the planet's coral reefs are at risk of degradation. The Global Coral Reef Monitoring Network has found that one-fifth of the world's reefs have been degraded beyond recognition. We are in a state of global warming. The sea temperatures are rising, glaciers melting and a new passage way has opened up above Alaska due to snow melting. Higher sea temperatures cause acidification and decreased dissolved oxygen. A century ago there were 150 glaciers in Glacier National Park in Montana-now 25 remain. Exotic plants and animals have been brought to us in the United States by ship bilge water as well as by human hand. Kudzu vines purposely planted to stabilize the soil now grow over houses, trees, anything in its path every summer in the south. Asian carp threaten the Great Lakes, exotic mussels clog pipelines, the Burmese Python now calls Florida home, and feral swine, nutria, ash borer beetles and more call the USA home. Here are some species that have gone extinct in the past 40 years.

The Golden Toad-pollution and skin infection-1989

The Zanzibar Leopard-last one shot in 1996

The Po'ouli Bird, Maui-loss of habitat

The Madeiran Butterfly-loss of habitat

Tecopa Pupfish, Mojave Desert-loss of habitat-1982

Pyrenean Ibex-last one shot in 2000

Black Rhinoceros, West Africa-hunted for its horn thought to be an aphrodisiac in 2006

There are more, but this gets the point across. There has been migration of fish and other animals of the ocean, to cooler water. The Lionfish with its stinging spines used to live in water close to the equator and now is living off the coast of North Carolina. People are being begged to catch them and eat them. Shells of oysters and other bivalves are thinner and less protective.

Not only are the air and water polluted, but also we have light pollution. Have you tried to see the stars and constellations recently? It's impossible anywhere near a city. There is also noise pollution. Enough said.

Seven billion and counting. We need another Earth. Famine and atomic leaks or wars face us, no doubt. How did we-Homo sapiens-with all our large brains let it get this way? Was it greed, speedier cars and more of them, selfishness or just not using our brains properly? It is true, there have been times of heat and times of cold as can be studied by taking cores from deep in the earth. So, are we then in a natural geologic cycle? Maybe our thoughtlessness didn't cause this mess after all. What, if anything, can we do about it? We continually fight each other over religion, space or political differences. History keeps right on repeating itself. Rats in a maze with increasing population end up killing each other. It has been suggested, and rightly so, that we be more compassionate toward each other, especially toward the land and animals that don't possess a "sapiens" brain. Couldn't we be better stewards of nature?

Rather than end on an uncheerful note, we shall examine some very unusual animals in unusual places.

The Dark and the Deep

Most of the ocean fauna live in the uppermost 500-1000 feet depending on the clarity of the sea water. Generally water near shores and mouths of rivers is somewhat cloudy due to particulate matter as well as plankton. The sun, then, will penetrate only 400-500 feet. Pelagic (outer) sea water has a much higher clarity allowing the sun rays to penetrate much further, to about 1000 feet. The creatures of the ocean floor that live by means of chemosynthesis have been discussed earlier. Until the mid-nineteenth century, it was thought there was no life below the waters lit by the Sun. Some individuals had tried sounding the Atlantic and at many points were unable to reach bottom. Scientists knew the sea water became colder in the deep and also that the pressure increased. It was, therefore, assumed that no life could exist in the deep ocean. In 1815, Edward Forbes, a medical student at Edinburgh, became interested in sea life. He began using dredges in deeper and deeper water and began hauling up creatures from deeper and deeper water. Creatures he could not identify. He knew a submarine was needed in order to identify these creatures, but unfortunately he died before an investigation was begun.

While trying to splice the ends of an underwater cable from the United States to England in 1866, the wire popped apart (more than once) and on close investigation it was found to have living coral attached to it. No eyebrows were raised. Later on the German biologist Ernst Haeckel examined some "mud" from the bottom of the ocean and thought it was alive. It was the original slime. He thought it to be partially protoplasm and named it Monera. This was in 1857. It was thought to be the lowest form of animal life. Later on this theory was disproved and thought it instead to be a colloidal substance. The sea floor was later dredged (1868) and new species of star fish, urchins and shellfish obtained. The Challenger was the ship from which dredges were put down and came up with more and more unusual life forms. Heads of State in Monaco were responsible for controlling the Atlantic Ocean originally. When it became clear to them that scientific study was called for, Monaco relinquished control to scientists. Prince Albert I never lost his interest in oceanography (a new term in the 20th century). On his ship, the Princess Alice, he approached a whaler that had just captured a sperm whale. The whale was regurgitating a large squid never before seen. Prince Albert remained interested in oceanography; he outlined the Gulf Stream in the Atlantic and subsidized the Institute of Oceanography in Paris.

Later on submersibles were built, one of the first being *Alvin* out of Wood's Hole, Massachusetts and study of the deep began in earnest. Many creatures have adapted to the non-photic life of the abyss and to its high hydrostatic pressures and near freezing temperatures. Many have been found and studied especially since the unmanned submersible was built. The early manned submersibles such as Alvin were dangerous and also limited in their ability to photograph and collect specimens. Evolution of deep water fauna is thought to have come from either the photic or sea floor. Animals may have wandered higher or deeper and adapted to those different conditions. We know there was no life in the Atlantic before the Cretaceous Period, 135 million years ago, because Africa and South America were just beginning to part. As for the Pacific, Indian and Southern Oceans, one can only suspect it was the wanderers and their capacity for adaptation that ruled the deep. The most common animal on the sea floor is the sea cucumber. They appear in odd forms and are brightly colored-orange, yellow, purple or violet. The brittle star is also seen frequently leaving their typical multi-legged trail. Fauna that live in the deep water are nearly free of gravity. Because water is 800 times more buoyant than air, marine creatures are free to change their shapes, thus appearing rather weird to us. Indeed, some are even difficult to put in the correct phylum. Almost all deep sea fauna have bioluminescence, either continuous or the ability to cut it on and off. In the ocean there is a continuous drift of "snow". It goes to the ocean floor and is made up of shed exoskeletons, dead phytoplankton, silica from diatom capsules and other amorphous material. It can be food for young fauna in the deep or drop into the ooze on the bottom. Salps are invertebrate filter feeders occurring in chains with mucous strands. They look like transparent cylinders and have been found at 1,400 feet perfectly healthy. Many species of glass sponges have been found rooted to the ocean floor. If the sponge tissue dies, the glass spicules linger on looking like an eerie glass forest. Many jellies live in the deep sea. The species are quite different from those in the photic zone. Most are medusas, purple or deep red and luminescent. The deep trenches, the hadal zone, are almost inaccessible. One trawl in a Pacific trench about 35,000 feet down contained polychaete (segmented) worms, anemones, sea cucumbers, crinoids, star fish and small fish and they were all blind. Unneeded eyes had "dis-evolved". Starfish (sea stars) in the deep sea lack podia or suction feet-unnecessary in buoyancy. A sea cucumber was found with a "sail" extending from its body. Another sea cucumber in the deep is rounded with spikes above its body. These are *Scatoplanes globosa* and occur in herds. Another deep sea cucumber lives on the floor and moved along on its podia. It also has protrusions from the top of its body, rendering a resemblance of *Wiwaxia* of the Cambrian! Many copepods live in all levels of depth in the ocean. They are that important in the food web. Deep sea shrimp are always scarlet. The color red has a short wave length and will appear black in the depths, giving the shrimp protection.

The evolution of deep sea fauna is unclear, but glypheid shrimp that were common in the Jurassic (213-144 million years ago) and thought to be extinct by the Eocene (50 million years ago) have been found off the coast of the Philippines and more recently from the Coral Sea Trawl at 1,329 feet and called the "Jurassic" shrimp but given the new genus. It was white with many red spots. Deep sea squid have long been objects of curious and perhaps deadly ambience. The "Vampire Squid from hell" certainly looks the part with its mantle lifted and its single eye glaring at you. It can live where oxygen levels are lower and its color is a dazzling scarlet. The octopus has adapted very well to living at 17,000 feet down. It has a web-like umbrel which helps it float along bobbing slightly. The deep sea fishes have evolved in such a way as to maximize its chances of finding and swallowing whole its prey. Most deep sea fishes are dark brown or black and they all have different assemblages of bioluminescence. The luminescence serves as a lure for prey and as a mate finder.

Vampire Squid

Angler fish

Black Swallower Eel

Eel Comb Jelly

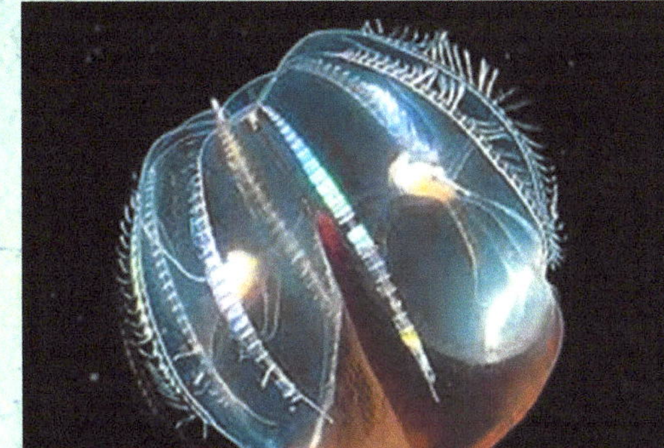

Strange Fish

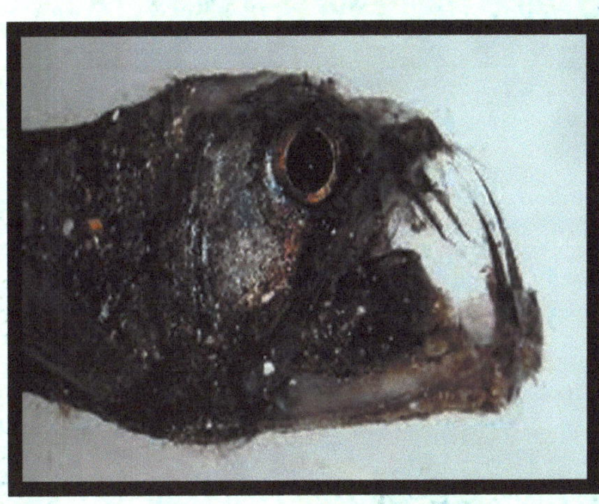

Dumbo Octopus

Deep Sea Viper

Some have bioluminescent symbiotic bacteria that produce a glow without the fishes control, but always an asset. One has a row of blue lights on each side. The "black swallower" is an eel six feet long with a tremendous low jawed mouth. It sports a luminescent lure at the terminus of its long, narrow obscure tail. Swimming at 8,000 feet by day and rising toward the surface at night is the terrible viperfish. Its lower jaw drops down in order to swallow fishes larger than it is. Quite a feat! The Angler fish has a luminescent lantern on its head and below its chin a long garden of branching filaments. This is the female. Her mate, the tiny male, lives attached to her as a parasite, but always there and ready! She may have as many as 6 males attached to her body for life. There are several with shining chin barbols and lantern lures above-all with long needle like teeth and mammoth mouths. We shouldn't be so hard on them. They live in pitch black darkness, unfathomable depths of water and like all life, the will to live is intrinsic.

Deep sea fauna are so well adapted to their harsh environment that they are difficult to study in the laboratory as their bodies quickly disintegrate. The deeper a fish lives, the more fragile their flesh and the more minimal their structure.

Lantern fish are the most numerous and diverse and the most widely distributed in the deep ocean. All deep sea fishes are slow to reproduce and become endangered from continental slope fishing.

Caves, caverns and grottos are formed by a combination of of geologic processes and many forms of life have found a home there. Bats use caves for rest and family life, leaving at night for feeding. Animals that live only in caves are called troglobites. They cannot live outside the cave and have adapted to life in darkness. Those fauna that live in complete darkness are often white in color and blind. Where there is water, fish, salamanders and shrimp can be found. Various insects and spiders spend their entire lives in the darkness of caves. There are numerous caves containing artistic work by humans. Cave paintings in Eurasia have been discussed. Cave paintings in Malaysia and South Africa have been dated from 60,000 years ago in the Paleogene.

There have been several events of mass extinction since life began on Earth. Before each of these extinctions, there is evidence of there having been a period of warming. Some scientists believe we may be warming now with the expectation of another mass extinction some time in the future. Some believe that loss of habitat plus rapid warming could cause an extinction of half the species in the next 100 years. Twenty-five percent of all mammal species may go extinct over the next 30 years. The World Global Census study found that 90% of all large fishes have disappeared in the past half century. Commercial use of fish and shellfish has caused a depletion not far from extinction. We may have passed the point of no return. The same holds true of life on land. Unless there is population control-if not voluntary, then famine, wars and uncontrolable disease may decrease the population and more green space could be then opened up. We need more policing of the oceans if we are to continue with enough oxygen on Earth, food on Earth. Our ocean populations must be protected and there must be less greed of Homo sapiens so that we can continue to wonder at the amazing life in the ocean and on our planet, Earth.

Bibliography

1. Dawkins, Richard, 2009-**The Greatest Show on Earth**, Simon and Schuster
2. Dunn, Rob, 2011-**The Wild Life of Our Bodies**, pages 4-14, Harper Collins
3. Ellis, Richard, 1996-**Deep Atlantic**, pages 24-25 and pages 218-297, Knopf
4. Fortey, Richard, 2000-**Trilobite!**, Knopf
5. Fortey, Richard, 1999-**Life, A Natural History of the First Four Billion Years of Life On Earth**, Random House
6. Fortey, Richard, 1982-**Fossils The History of Life**, Sterling
7. Gould, Stephen Jay, 1989-**Wonderful Life**, Norton
8. Gould, Stephen Jay, 2002-**I Have Landed**, pages 181-191 Harmony, Random House
9. Gould, Stephen Jay, 1995-**Dinosaur In a Haystack**, pages 221-237, Crown Publishers
10. Gould, Stephen Jay, 1985-**The Flamingo's Smile**, pages 197-198, Norton
11. Gould, Stephen Jay, et.al. 2001-**The Book of Life**, Norton
12. Kolbert, Elizabeth, 2011-**The Acid Sea**, National Geographic Volume 219 #4, pages 100-121
13. Lane, Nick, 2009-**Life Ascending The Ten Great Inventions of Evolution**, Norton
14. Latimer, Margorie Cortenay, **Ancient Swimmers**, National Geographic March 2011, pages 86-93
15. Margulis, Lynn and Sagan, Dorian, 1995-**What is Life?**, Simon and Schuster
16. Nouvian, Claire, 2007-**The Deep**, University of Chicago Press
17. Safina, Carl, 2011-**The View From Lazy Point**, pages 313-317, Holt & Company
18. Safina, Carl, 2011-**A Sea in Flames**, Crown Publishers
19. Shipman, Pat, 2011-**The Animal Connection**, Norton

20. Shubin, Neil, 2008-**Your Inner Fish**, Pantheon Books, Random House

21. Tester, Pat, 2011-NOAA Beaufort Lab, microbe eating oil re: Gulf of Mexico Spill, personal communication

22. Tester, Pat, 2011-NOAA Beaufort Lab, dinoflagellate evolution, personal communication

23. Tudge, Colion, 2000-**The Variety of Life**, pages 492-513, Oxford University Press

24. Weinberg, Samantha, 2000, **A Fish Caught in Time**, The Search for the Coelacanth, Harper Collins

25. Many authors, 2009-**Prehistoric Life**, DK Publishing

26. Zimmer, Carl, 2001-**Evolution The Triumph of an Idea**, Harper Collins

27. **World Ocean Census**, 2009, Firefly Books Ltd.

28. Wikipedia-free-information

29. Wikipedia-free-illustrations

30. Barnes, Ruppert-**Invertebrate Biology**, sixth edition, Harcourt Brace

Best Books for Illustrations

1. Fortey, Richard-**Fossils**
2. Gould, Stephen Jay- **The Book of Life**
3. Nouvian, Claire-**The Deep**
4. Shuford, Neil-**Your Inner Fish**
5. Multiple Authors-**Prehistoric Life**
6. Crist, Scowcroft & Harding-**World Ocean Census**

Glossary

apogee- The point in an orbit most distant from the body being orbited

Batholith- a very large irregular-shaped mass of igneous rock, esp. granite, formed from an intrusion of magma at great depth, esp. one exposed after erosion of less resistant overlying rocks

biramous-having two branches such as the appendages of crustaceans

bivalve-a marine invertebrate that has 2 shells such as a clam

byssal- A mass of strong, silky filaments by which certain bivalve mollusks, such as mussels, attach themselves to rocks and other fixed surfaces.

calcite-crystallized calcium carbonate

carnivorous-feeding on flesh

chemosynthesis-the chemical process whereby sulfur or methane produces sugars without light

chlorophyll-a green photosynthesis pigment

chordate-an animal with a notochord or vertebrae

convergent evolution-exhibiting convergence in form and characteristics of bodily structure though separated by distance

Coralline algae-red algae

Cycad-gymnosperms with pinnate leaves, little xylem and a thick cortex

extant-living today

fauna-animal life

flagella-an appendage to a cell which causes movement

flora-plant life

flotsam-floating wreckage or debris

fossil-preserved remnant of a long dead plant or animal

fusion-a chemical reaction in which there is merging of two elements into a new element. Example-2 hydrogen atoms fuse into one helium atom

gonads-sex organs

gorgonian Coral-a type of coral with a central axial rod and having a horny or calcareous branching skeleton

herbivorous-plant eating

iridium- a very hard inert yellowish-white transition element that is the most corrosion-resistant metal known

lycopod- primitive evergreen moss-like plant with spores in club-shaped strobiles

methane-a colorless, odorless flammable gas, the simplest alkane and the main constituent of natural gas: used as a fuel. Formula: CH_4 is a gaseous product of decomposition or can occur naturally

nutrients- any substances that nourish an organism

omnivorous- eating both animal and vegetable foods

organelles-inclusions in the protoplasm of a cell. An example is a food vacuole

ozone-an unstable, poisonous allotrope of oxygen, O3, that is formed naturally in the ozone layer from atmospheric oxygen by electric discharge or exposure to ultraviolet radiation, also produced in the lower atmosphere by the photochemical reaction of certain pollutants

paleontology-the study of the life of geological periods as known from fossil remains

perigee-t he point in any orbit nearest to the body being orbited

phloem- A tissue in vascular plants that conducts food from the leaves and other photosynthetic tissues to other plant parts

photosynthesis-the process by which carbon dioxide and water, stimulated by the sun makes sugars and oxygen

phylum-a group of primary divisions of the animal kingdom

plankton-a group of phyto plankton (one celled algae) and zooplankton (animals) that live in the uppermost layers of the ocean and is the first step or the basis for the food chain

pollenia-compressed, slightly sticky pollen occurring naturally on orchids

prokaryote-a cell without a nucleus such as a bacterium

protist-any of a group of unicellular organisms-one celled with nucleus and can exist in groups such as sponges

serrated-notched or toothed on the edge

syrinx- The vocal organ of a bird, consisting of thin vibrating muscles at or close to the division of the trachea into the bronchi

velociraptor- A small, fast, carnivorous dinosaur of the genus Velociraptor of the Cretaceous Period that was about (6.5 ft) in length. It had long curved claws for grasping and tearing at prey, walked on two legs that were adapted for leaping, and had a long stiff tail used as a counterweight.

xylem- a plant tissue that conducts water and mineral salts from the roots to all other parts, provides mechanical support, and forms the wood of trees and shrubs

Special Thanks

Many thanks to Lainy Vincent, my assistant. She chose all the pictures and made good suggestions on syntax and inclusions. I could not have written it without her.

Thanks also to Ricardo Galbis, M.D., my classmate in med school for sparking the idea and many thanks to B. L. Rish, M.D., for his encouragement and suggestions.

And thanks also to my husband, Thomas L. Gwynn, M.D. for his patience and good humor.

Cover microphotograph by Megan Black

Index

A

ACGT (adenine, cytosine, guanine, and thymine), 9
Afar Desert, 72
Africa, 37, 42, 67, 69
  geological movements of, 48, 53, 68, 77
  primates in, 64–65, 70, 73
Alaska, 74
Albert I, 76
Alvarez, Walter, 65
Alvin, 77
America, 68
  animals in, 48, 65, 68
  climate change in, 68
  geological movements of, 53, 59, 68, 77
ammonites, 28, 37–39, 50, 53
  origin of, 23
amphibians, 34, 37–38, 49, 53
amphioxus, 33
Andrewsarchus, 63
angiosperms, 59–60, 62
Angraecum sespidale, 10, 60
Annelida, 45
Anning, Mary, 54
Anomalocaris, 19
anteaters, 64
Anthopocene, 73
archaea, 9, 11
Arctodus, 68
Ardipithecus ramidus, 72
Aristotle's lantern, 22
arthropod, 29, 33, 45
Asia, 64–65, 67–69, 72
atmosphere, 9, 14, 25, 28–29, 47
  composition of, 8
aurochs, 68
Australia, 9, 42, 48
  animals in, 61, 68
  geological movement of, 48, 59
Australopithecus afarensis, 70, 72
Australopithecus sediba, 72

B

bacteria, sulfur, 10
bats, 64
Bering Sea, 68
Bering Strait, 42, 65
big bang theory, 8, 17
bilateralism, 17, 22
bioluminescence, 21, 77–78
bivalves, 28, 39, 50
brachiopods, 29, 34, 38–39, 50, 53
  characteristics of, 23, 27
  development of, 30
  origin of, 22–23
Brachiosaurus, 55, 58
Brassia, 60
Briggs, Derek, 18
brittle star, 77
bryophytes, 18
bryozoans, 27–28, 34, 39
Burgess Shale, 18, 21–22, 24, 26

C

Calcutta, 74
Cambrian, 15–18, 21–22, 25, 28, 44, 78
  beginning of certain life-forms during, 18, 22–23, 28, 33, 44–45, 47
  biological evolution during, 17
Canada, 18
Canis dirus, 68
carbon, 14, 16, 68
carbon dating, 16
Carboniferous, 37–40, 59, 74
  climate changes during, 38–39
Carboniferous Rainforest Collapse, 38
cattle, 68
Cave of Forgotten Dreams, 73
cave paintings, 73, 80
Cenozoic, 62, 68
cephalopods, 60
chemosynthesis, 8–10, 76
China, 64
chloroplasts, 12, 14, 30
choanoflagellates, 15
chordates, 23, 46
chromataphores, 44–45
cirripedia, 45
cladistics, 34

Cladocora of Cyprus, 66
cloning, 44
coal, 37–38, 74
coelacanth, 33
coevolution, 60
cold seeps, 11
conifers, 39, 51, 56, 62
conodont, 26
continents, 33, 51, 59, 68–69, 73
  division of, 53
  fossils in the, 58
  gyres circling the, 42
cooksonia, 31–32
copepods, 13–14, 45, 78
coralline algae, 53
coral reefs, 22, 28, 34, 53, 62
  characteristics of, 22, 27, 29
  formation of, 22
  introduction of new species in, 28
Coral Sea Trawl, 78
Crepidula, 44
Cretaceous, 53, 59, 77
  animals in the, 57, 61
  plants in the, 59–60
Crichton, Michael, 56–57
Crick, Francis, 9
crinoids, 31, 34, 37–39, 45, 50, 77
  characteristics of, 30
  first appearance of, 22
crocodiles, 54
Cro-Magnons, 72
crustaceans, 53
ctenophore, 21
currents, types of, 42
cuttlefish, 44–45
cyanobacteria, 9, 16
cycads, 39
Cycle, Calvin, 14
cynodonts, 50–51

D

Darwin, Charles, 9–10, 17–18, 46
Darwin's orchid. *See Angraecum sespidale*
Deinotherium, 66
Devonian, 29, 33–36, 43
diatoms, 14, 29
diatryma, 62
Dimetrodon, 40
Dimorphodon, 54
dinoflagellate, 29–30
dinosaurs, 48, 52, 54, 57–59, 62
  characteristics of, 55, 58–59
  extinction of, 23, 51, 59, 62, 65
  media relating to, 56–57
DNA (deoxyribonucleic acid), 9, 46, 65, 70
domestication of animals, 69
Dryopithecus, 67

E

echidna, 61
echinoderms, 13, 21–22, 30–31, 38, 45
Ecuador, 11
Ediacaran, 6, 12, 15
Elasmosaurus, 60
Elkinsia, 36
eocaecilia, 53
Eocene, 6, 62–65, 69, 78
Eomaia, 61
Ethiopia, 72
eudimorphodon, 49
euparkeria, 49
Eurasia, 59, 65, 68, 80
extinct species, 74

F

ferns, 36, 39, 51, 56, 62
filter feeders, 25, 27, 40, 43
foraminifera, 38
Forbes, Edward, 76
France, 72–73
fungi, 11

G

Galapagos, 10
gastropods, 28, 50
Germany, 49, 73
Gibbons, Ann, 72
ginkgo, 40, 51, 56
glaciation, 28, 66, 68
Glacier National Park, 74
Global Coral Reef Monitoring Network, 74
Glossopteris, 39
Gonawanda, 53
Gondwandaland, 37

Gould, Stephen J., 18, 25
Wonderful Life, 18
graptolites, 25
Great Barrier Reef, 28
Greenland, 34, 42, 49, 66
Gulf of Mexico, 43

H

Haeckel, Ernst, 76
hagfish, 25, 31
Harzog, Werner, 73
Hawking, Stephen, 8
hermaphrodites, 44
Herrerasaurus, 49
homeotherms, 46
hominids, 65, 70–72
Homo erectus, 69, 72
Homo ergaster, 71–72
Homo habilis, 70–71
Homo neanderthalensis, 72
Homo sapiens, 69, 71, 73, 81
horses, 68, 73
horseshoe crabs, 19–20, 45
horsetails, 34, 56
human revolution, 72–73
Hybodus, 53

I

ichthyostega, 34
Icthyosaurus, 54
insects, 61, 64
colony, 60
flying, 37, 52
Institute of Oceanography in Paris, 77
invertebrates, 34, 43, 53, 60, 62, 65
iridium, 59, 65
Isthmus of Panama, 66

J

Java, 72
jellyfish, 15, 17–18, 21
Jurassic, 53–57, 68, 78
Jurassic Park, 56–57

K

Kenya, 72
Komodo dragon, 68

L

Laurasia, 53, 69
leeches, 29
lemurs, 69
Lepidotes, 53
Leptomeryx, 64
Limulus, 45
Linnaeus system of binomial nomenclature, 25
liverworts, 18
Luciferan, 21
Lucy. *See Australopithecus afarensis*
lungfish, 34
lycopods, 37

M

Madagascar, 69
Malaysia, 80
marmoset, 64
Marrella, 18
marsupials, 48, 61, 70
Mastodon, 66
Mesozoic, 6, 48, 69
Miocene, 65–66, 69
mitosis, 22
molecular biology, 34
mollusks, 13, 23, 29, 44
Monaco, 76
Monera, 76
monkeys, 69
Morris, Simon, 18
Moschops, 40
mosses, 18, 31–32, 34
mountain ranges, 66

N

natural selection, 9–10, 46–47
nautiloids, 50
nautilus, 23, 28, 44
nematocysts, 15
Neogene, 66

notochord, 23–24, 33, 46, 70

O

oceans, 8, 11, 20, 42, 66, 68, 76–77
  animals in, 11, 20
  geological formation of, 23, 33, 53, 59, 77
octopus, 44, 78
Old Faithful, 11
Oligocene, 62
Oliva, 66
opossum, 48
orchids, 10, 60
Ordovician, 25–29, 31
oxygen, 9, 14
ozone layer, 8

P

Pacific trench, 77
Paleocene, 62
Paleozoic, 22, 26, 39, 45, 48
Pangaea, 39, 48, 53, 59
parasites, 29, 80
Parasuchus, 49
penguin, 65
Permian, 20, 22, 39–40, 51, 73
  animals during the, 40
  duration of the, 39
  evolution during the, 40
  extinction during the, 20, 39, 41, 47–48, 59
  landscape during the, 41
Philippines, 78
photosynthesis, 9, 29, 35, 62
  first life capable of, 9
  process of, 8, 10, 14, 22, 32
phylum, 17–18, 43–46
phytoplankton, 12, 14, 29
Pikaia, 24, 26, 33
plankton, 12–14, 29, 43, 45–46, 50, 76
plants, carnivorous, 60
Plateosaurus, 49
platypus, 51, 61, 70
Pleistocene, 68
pleuromeia, 51
Pliocene, 65–66, 69
Pliohippus, 66
pollution, 14, 37, 43, 74–75
polyps, 22
primates, 64, 69–71
protists, 12
Protoceratops, 59
Protohippus, 64
Prototaxites logani, 35
Pteranodon, 61
Pterodactylus, 54
pterosaurs, 48–49, 54–55, 61

R

radiolarians, 38
Raurisuchia, 48
RNA (ribonucleic acid), 9
rodents, 65

S

saber-toothed tiger, 68
salps, 77
Sanctacaris, 19
sand dollar, 22
scallops, 23
sea cucumbers, 22, 45, 77
sea horses, 44
sea stars, 22
sea urchins, 22, 27, 45
Selaginella, 56
sharks, 34, 38, 40, 60
shellfish, 76, 81
shrimp
  deep-sea, 78
  glypheid, 78
Silurian, 23, 25, 29, 31–33, 36
smokers, 10
South Africa, 49, 72, 80
southern hemisphere, 39–40, 42
Spain, 73
spectrometry, 16
spicules, 15, 44
Spielberg, Steven, 56
sponges, 15, 17–18, 21, 28, 33, 43
squids, 34
starfish, 77
Stegosaurus, 55
stomata, 32
stony coral polyps, 22, 27, 30, 44
stromatolites, 9
sugar shuffle, 14

sun, 8, 10, 46, 60, 62, 76
symbionts, 22

T

tabulate corals, 30
Tertiary, 62, 65
tetrapods, 33, 37, 40, 46, 48
thelodont, 31
Therapsids, 50
theropod, 54
Tiktaalik, 33–34
trilobites, 17–18, 25, 27, 29, 34
  characteristics of, 20–21, 45
  extinction of, 39
triploblastic life, 17
turtles, 37, 43, 54, 58, 63
Tyrannosaurus Rex, 49, 58

U

ungulates, 65

V

Varanops, 40
Velociraptor, 57, 59, 86
vertebrae, 26, 33–34, 40, 47, 60
  animals with, 27, 33, 48, 53
  development of, 24
viperfish, 80

W

Walcott, Charles, 18
Wallace, Alfred Russel, 18
Watson, James D., 9
Weltrichia, 56
whales, 64
White, Tim, 72
Whittington, Harry, 18
Williamsonia, 56
Wiwaxia, 19, 78
Wonderful Life (Gould), 18
World Global Census, 80
worms, 27–28, 44–45, 77
  velvet, 28
Wyoming, 63–64

Y

Yellowstone National Park, 9, 11

Z

zooplankton, 13, 29
zooxanthellae, 22, 30

www.ingramcontent.com/pod-product-compliance
Ingram Content Group UK Ltd.
Pitfield, Milton Keynes, MK11 3LW, UK
UKHW060120300726
14090UKWH00002B/280
* 9 7 8 1 4 6 5 3 7 8 5 8 3 *